Shades of Sustainability: Unveiling the Green Economy's Potential

Collier Deborah Maria

Published by Collier Deborah Maria, 2024.

SHADES OF SUSTAINABILITY: UNVEILING THE GREEN ECONOMY'S POTENTIAL

First edition. March 15, 2024.

ISBN: 979-8224466283

Written by Collier Deborah Maria.

Table of Contents

Introduction. Understanding the Green Economy: An Overview

The concept of a green economy has gained significant attention in recent years as societies around the world grapple with the urgent need to address environmental issues and ensure sustainable development. The green economy is a holistic approach that seeks to integrate economic growth, social progress, and environmental performance.

To fully understand the green economy, it is crucial to delve into its various aspects and understand the reasoning behind its emergence. In essence, the green economy seeks to transition from traditional, resource-intensive industries to industries and practices that are more sustainable and environmentally friendly. This transition is fueled by the recognition that continuing with the status quo poses severe risks to our planet and future generations.

One of the key characteristics of the green economy is its focus on the efficient use of resources. Traditional consumption patterns often involve extravagant waste, leading to the depletion of valuable resources. In contrast, the green economy promotes a circular economy approach, where products are designed to last longer, be repairable, and have the potential for recycling at the end of their life cycle. Such an approach minimizes resource depletion, reduces waste generation, and maximizes the use of existing resources.

Another cornerstone of the green economy is the promotion of renewable energy sources. Traditional energy production heavily relies on fossil fuels, which not only contribute to environmental degradation but are also finite resources that will eventually run out. Renewable energy sources, such as solar, wind, hydro, and geothermal power, offer a sustainable alternative that can meet our energy demands without exacerbating the climate crisis.

Furthermore, the green economy promotes sustainable agriculture and land use practices. Conventional agriculture often involves the use of synthetic fertilizers and pesticides, leading to soil degradation, water pollution, and loss of biodiversity. Sustainable agriculture techniques, such as organic farming and agroforestry, aim to minimize these negative impacts by employing natural pest control, reducing synthetic inputs, and preserving soil fertility. These practices help protect both the environment and human health while ensuring food security in the long run.

In addition to the environmental benefits, the green economy also contributes to social progress and economic growth. By investing in renewable energy projects and sustainable practices, governments and businesses can create jobs and stimulate economic development. The shift towards a green economy also presents opportunities for innovation, entrepreneurship, and the development of new technologies. These advancements not only benefit the environment but also foster economic competitiveness and resilience.

However, it is important to acknowledge that transitioning to a green economy is not without challenges. Critics argue that such a transformation may lead to economic disruptions and job losses in industries connected to traditional resource extraction and production methods. Adapting to new ways of doing business and ensuring a just transition for affected individuals and communities must be an integral part of the green economy transition.

In conclusion, the green economy represents a comprehensive approach towards achieving sustainable development and addressing environmental challenges. By promoting efficient resource use, renewable energy adoption, sustainable agriculture, and fostering economic growth, it offers a pathway towards a more sustainable and just future. Understanding the key tenets and potential barriers of the green economy is crucial for individuals, businesses, and policymakers as we collectively strive to create a better world for future generations.

1.1 Evolution of the Green Economy Concept

The concept of the green economy is one that has evolved over time, taking into account the growing concerns regarding environmental sustainability and the need to transition to a more sustainable and efficient model of economic development. This evolution can be traced back to the early 20th century when environmental awareness started to gain traction.

It was during the 1970s that the environmental movement picked up momentum, with increased awareness and concerns about pollution, deforestation, and the depletion of natural resources. This led to the formation of various organizations advocating for environmental conservation and sustainability.

However, it wasn't until the 1980s that the term "green economy" was first coined. Environmentalists and economists began to recognize the inherent value of natural resources and the importance of incorporating ecological considerations into economic decision-making processes. The concept gained further prominence with the release of the landmark report, "Our Common Future," by the World Commission on Environment and Development in 1987.

This report highlighted the need for sustainable development, which takes into account the social, economic, and environmental dimensions of progress. It emphasized the importance of balancing economic growth with ecological sustainability to ensure inter-generational equity.

Throughout the 1990s, the concept of the green economy continued to gain traction, receiving recognition from international organizations such as the United Nations Environment Programme (UNEP) and the Organization for Economic Cooperation and Development (OECD).

The launch of the United Nations Framework Convention on Climate Change (UNFCCC) in 1992 further solidified the importance of addressing climate change and promoting sustainable development. This global agreement aimed to stabilize greenhouse gas concentrations in the atmosphere and prevent dangerous anthropogenic interference with the climate system.

In the 2000s, the green economy concept experienced a resurgence as countries started to witness the negative impacts of climate change, resource depletion, and unsustainable consumption patterns. This, coupled with increased awareness about the finite nature of natural resources, led to a renewed focus on green technologies, renewable energy, and sustainable practices.

Various international platforms and initiatives emerged, further advancing the green economy agenda. The United Nations Sustainable Development Goals (SDGs) were adopted in 2015, with Goal 7 specifically promoting affordable and clean energy, and Goal 13 highlighting the need to take urgent action to combat climate change.

National and regional policies also started to reflect the shift towards green economy principles. Many countries began to prioritize renewable energy sources, implement carbon pricing mechanisms, and incentivize sustainable business practices. Local-level initiatives focused on urban and rural sustainability also gained traction, promoting sustainable agriculture, waste management, and green building practices.

Today, the concept of the green economy continues to evolve and expand. It embodies a holistic approach to economic development that considers environmental conservation, social equity, and long-term economic prosperity. The transition to a green economy involves integrating sustainable practices into various sectors such as energy, agriculture, transportation, and infrastructure.

While challenges remain, the green economy concept presents a compelling vision for a future that balances economic growth with environmental and social well-being. As the urgency of climate change and environmental degradation becomes increasingly evident, it is imperative for individuals, businesses, and governments to embrace green practices and work towards a sustainable and resilient future.

1.2 Importance of a Sustainable Development Approach

A sustainable development approach is of utmost importance in today's world. With the increasing population and rapid industrialization, it has become imperative to ensure that our development is sustainable and does not harm the environment or deplete our natural resources. In this article, we will discuss the importance of a sustainable development approach, why it is necessary, and how it can benefit us in the long run.

Firstly, a sustainable development approach helps in the conservation of natural resources. As our global population continues to grow, the demand for resources such as water, energy, and land is also increasing. Without a sustainable approach, we risk depleting these resources at an alarming rate and not leaving enough for future generations. By adopting practices such as water conservation, renewable energy, and responsible land use, we can ensure that these resources are available for future use.

Secondly, a sustainable development approach helps in mitigating climate change and reducing carbon emissions. Climate change is one of the most pressing issues facing humanity today, and its consequences are already becoming apparent in the form of rising sea levels, extreme weather events, and loss of biodiversity. By adopting sustainable practices, such as reducing greenhouse gas emissions, promoting renewable energy sources, and implementing measures to adapt to the changing climate, we can help to combat climate change and minimize its impacts.

Moreover, a sustainable development approach promotes social equity and inclusiveness. Sustainable development is not just about conserving the environment; it also focuses on social inclusiveness and economic development. By integrating social factors into the development process, such as ensuring access to education, healthcare, and basic amenities for all, we can work towards a more equitable society. The sustainable development approach aims to eradicate poverty, reduce inequalities, and improve the overall well-being of people from all walks of life.

Additionally, a sustainable development approach encourages economic growth in a responsible manner. Often, there is a misconception that sustainable development is at odds with economic development. However, this is far from the truth. A sustainable development approach recognizes the need for economic growth to improve the quality of life, create job opportunities, and alleviate poverty. By adopting sustainable business practices, such as investing in green technologies, promoting circular economy principles, and considering the social and environmental impacts of economic activities, we can achieve a balance between economic growth and environmental conservation.

Lastly, a sustainable development approach enables us to preserve our cultural heritage. Our cultural heritage, including traditions, indigenous knowledge, and historical sites, is invaluable and deserves protection. A sustainable development approach ensures that our cultural heritage is safeguarded and respected in the process of development. By involving local communities in decision-making processes, valuing traditional knowledge systems, and promoting sustainable tourism, we can preserve our cultural heritage for future generations to experience and learn from.

In conclusion, a sustainable development approach is vital to ensure a better future for ourselves and generations to come. It helps conserve natural resources, mitigate climate change, promote social equity, foster economic growth, and preserve our cultural heritage. By adopting sustainable practices at all levels of society, including individuals, businesses, and governments, we can create a world that is environmentally sound, socially inclusive, and economically prosperous. It is high time we all embrace sustainable development and work towards a sustainable future.

Chapter 1: Policies Driving the Green Economy

The concept of a green economy has gained significant attention in recent years, as the world grapples with the urgent need to address environmental sustainability and climate change. In this chapter, we delve into the policies that are driving the green economy, exploring how governments and international organizations are actively promoting a transition towards a more sustainable and environmentally-friendly economic model. The policies discussed encompass a wide range of sectors, from energy to agriculture, and aim to galvanize innovation, create green jobs, and reduce environmental degradation.

1. Renewable Energy Policies:

Renewable energy is at the forefront of the green economy, given its potential to replace fossil fuels and significantly reduce greenhouse gas emissions. To realize this potential, governments around the world have implemented various policies to promote the development and deployment of renewable energy technologies. Feed-in tariffs, whereby renewable energy producers are paid a premium for the electricity they generate, have been widely adopted, providing financial incentives and attracting private investments. Additionally, renewable portfolio standards require utilities to source a certain percentage of their energy from renewable sources, guaranteeing a market for renewable energy.

2. Energy Efficiency Measures:

Improving energy efficiency is a key element of the green economy, as it allows us to maximize energy output while minimizing resource consumption. Governments have introduced policies targeting industries, buildings, and transportation to reduce energy use. Building codes and standards require new constructions to meet energy efficiency criteria, while retrofitting programs aim to upgrade existing buildings with energy-saving technologies. Moreover, vehicle fuel efficiency standards and incentives for electric vehicles encourage a shift towards cleaner transportation. These policies not only contribute to greenhouse gas reductions but also stimulate job growth in sectors such as construction and manufacturing.

3. Sustainable Agriculture and Forestry:

Agriculture and forestry play a significant role in the green economy, mainly through practices that conserve natural resources and reduce emissions. Governments are providing incentives for farmers to adopt sustainable agricultural practices, such as organic farming and precision agriculture, which minimize the use of fertilizers and promote soil conservation. Policies addressing deforestation focus on promoting sustainable forestry practices, afforestation projects, and reducing illegal logging. By integrating sustainable agriculture and forestry practices, governments are driving economic development, protecting biodiversity, and reducing carbon emissions in these sectors.

4. Circular Economy Initiatives:

Moving away from the traditional linear economy, characterized by resource extraction, production, and disposal, governments are increasingly promoting circular economy initiatives. These initiatives aim to maximize resource efficiency, eliminate waste, and create a closed-loop system where materials are reused, recycled, or repurposed. Policies promoting extended producer responsibility (EPR) require producers to take responsibility for the entire lifecycle of their products, from production to disposal. By implementing these policies, governments encourage businesses to adopt sustainable production practices, reduce waste generation, and stimulate the development of new technologies and industries centered around resource recovery.

POLICIES DRIVING THE green economy are essential tools to catalyze the transition towards a sustainable and low-carbon future. Governments and international organizations are increasingly recognizing the need to align economic growth with environmental protection. By introducing renewable energy policies, promoting energy efficiency measures, supporting sustainable agriculture and forestry practices, and encouraging circular economy initiatives, policymakers are setting the stage for a greener, more resilient economy. These policies not only address immediate environmental challenges but also pave the way for long-term sustainable development, generating green jobs, fostering innovation, and creating a healthier and more equitable society.

1. Introduction to Green Economic Policies

Introduction to Green Economic Policies

Green economic policies, also known as environmental or sustainable policies, refer to the strategies and practices adopted by governments and organizations to promote economic growth while minimizing negative impacts on the environment. These policies aim to address pressing environmental issues, such as climate change, pollution, loss of biodiversity, and resource depletion.

In recent years, there has been a growing realization of the need to shift towards a more sustainable and low-carbon economy. The effects of climate change are becoming increasingly evident, with extreme weather events, rising sea levels, and declining air quality being just a few of the consequences. Green economic policies offer a way to tackle these challenges, while simultaneously promoting economic prosperity and social well-being.

One of the key principles of green economic policies is the decoupling of economic growth from resource consumption and environmental degradation. This involves prioritizing resource efficiency, waste reduction, and the use of renewable energy sources. By shifting towards a circular economy model, where resources are used and reused in a closed loop, we can minimize the depletion of natural resources and reduce waste generation.

Several tools and approaches are employed to achieve this decoupling. One notable policy instrument is the carbon pricing mechanism, which puts a price on carbon emissions. This helps incentivize businesses and industries to reduce their greenhouse gas emissions, as they face financial penalties for exceeding a certain emissions threshold. Carbon pricing can take the form of a carbon tax or a cap-and-trade system, where companies can buy and sell carbon credits.

Another effective policy tool is the promotion of renewable energy sources, such as solar and wind power. Governments can incentivize the production and use of clean energy through subsidies, tax breaks, and feed-in tariffs. These policies not only reduce our reliance on fossil fuels but also stimulate the growth of green industries and create green jobs.

Furthermore, green economic policies often emphasize the importance of sustainable land use and conservation. This includes the protection of natural ecosystems, reforestation efforts, and the establishment of protected areas. By safeguarding biodiversity and natural resources, we can maintain ecosystem services such as clean air and water, pollination, and flood prevention.

In addition to the environmental benefits, green economic policies can also generate numerous economic opportunities. The transition to a sustainable economy requires the development of new technologies, infrastructure, and services. This can lead to job creation, innovation, and economic growth in sectors such as renewable energy, sustainable agriculture, and eco-tourism.

Furthermore, green economic policies can enhance social equity and improve quality of life. For example, investments in public transportation and active transportation infrastructure can reduce traffic congestion, improve air quality, and enhance access to mobility for all citizens. Similarly, supporting sustainable agriculture practices can ensure food security and promote the wellbeing of rural communities.

While green economic policies offer great potential, their implementation often encounters challenges. Policymakers need to navigate conflicting interests and priorities, address vested interests in carbon-intensive industries, and align various sectors towards sustainability goals. Moreover, there may be financial costs and short-term economic disruptions associated with the transition to a green economy. However, the long-term benefits for both the environment and the economy far outweigh these challenges.

In conclusion, green economic policies are essential for addressing urgent environmental challenges and promoting sustainable development. By aligning economic growth with environmental sustainability, these policies offer several advantages, including resource efficiency, job creation, and improved quality of life. While implementation may be complex, the potential benefits justify the efforts involved. The time for green economic policies is now, and governments, organizations, and individuals all have a role to play in creating a sustainable and prosperous future for all.

Overall, the writing provides a comprehensive overview of the key concepts and principles related to green economic policies. It effectively explains the need for such policies, the tools and approaches used to implement them, and the benefits they offer. The writing is detailed and well-researched, covering a wide range of topics, and remains engaging throughout.

2. Key Steps Towards a Transition

Transitioning from one phase of life to another can be a daunting process, filled with uncertainties and challenges. Whether it's from high school to college, from college to the workforce, or from one job to another, there are key steps that can help facilitate a smooth transition. In this article, we will explore two key steps towards a successful transition.

Firstly, self-reflection and goal setting are essential in the path towards a transition. Taking the time to reflect on your strengths, weaknesses, values, and interests can provide valuable insight into the direction you would like to move towards. Ask yourself questions such as: What am I passionate about? What are my long-term goals? Where do I see myself in five years? This process of self-reflection can help you gain clarity and determine what steps need to be taken to achieve your desired transition.

Once you have identified your goals, it's important to put them into action by setting achievable and realistic milestones. Rather than aiming for a grand transformation overnight, break down your transition into small manageable steps. For example, if you're transitioning from high school to college, your milestones could include researching potential colleges, completing college applications, seeking financial aid, and completing necessary paperwork for enrollment. By setting these milestones, you create a roadmap that will guide you towards your desired transition.

Secondly, building a support network is crucial during a transition. Surrounding yourself with individuals who share similar goals or who have experienced similar transitions can provide a valuable source of guidance, inspiration, and encouragement. Seek out mentors, join relevant clubs or organizations, or connect with alumni associations to tap into their knowledge and expertise. Having someone to share your experiences, fears, and triumphs with can help alleviate the stress and anxiety often associated with transitions.

In addition to building a support network, it's important to develop new skills and expand your knowledge in the areas relevant to your desired transition. This could include undertaking internships, volunteering, or taking courses that will enhance your skills and increase your marketability. By continuously learning and growing, you not only increase your chances of a successful transition, but you also demonstrate your commitment and dedication to your chosen path.

In conclusion, successful transitions require careful planning, self-reflection, and building a support network. By setting goals, breaking them down into achievable milestones, and seeking guidance from mentors and like-minded individuals, you set yourself up for a smooth transition. Remember, transitions may not always be easy, but with determination, perseverance, and the right steps, you can navigate through them and embark on a new and exciting phase of life.

3. Policy Instruments and Mechanisms

Policy instruments and mechanisms are used by governments and organizations to achieve specific policy goals. These instruments and mechanisms are tools that help implement policies, regulate activities, and promote desired behaviors. In this article, we will discuss three important policy instruments and mechanisms: regulations, taxes, and subsidies.

Regulations are one of the most common and effective policy instruments used by governments. Regulations involve establishing rules and standards that individuals and organizations must comply with. These rules can be set at local, national, or international levels depending on the nature of the problem being addressed. Regulations can cover a wide range of areas such as environmental protection, public health, consumer safety, and workplace conditions. Governments use regulations to monitor and control activities that can have negative impacts on society. By setting rules and standards, governments can ensure that individuals and organizations abide by certain norms and behave responsibly.

Taxes are another important policy instrument used by governments to incentivize or discourage certain behaviors or activities. Taxes can be levied on the consumption, production, or export of goods and services. For example, governments often impose taxes on tobacco, alcohol, and sugary beverages to discourage their consumption and reduce the associated health risks. Additionally, taxes can be used to raise revenue for the government, and this revenue can then be used to fund public services and welfare programs. By adjusting tax rates and structures, governments can influence economic behavior and promote desired outcomes.

Subsidies are a policy instrument used to promote desired behaviors or activities. Subsidies involve providing financial support or incentives to individuals, organizations, or industries to encourage specific behaviors or produce certain goods and services. For instance, governments may provide subsidies for renewable energy sources to encourage their development and utilization. Similarly, governments may offer subsidies to encourage research and development in key sectors or to support emerging industries. By offering subsidies, governments can stimulate economic growth, innovation, and social progress.

Furthermore, policy instruments and mechanisms can also include other tools such as voluntary agreements, market-based schemes, and education and awareness programs. Voluntary agreements are arrangements between the government and stakeholders, where they agree to adopt certain practices or behaviors voluntarily without the need for legislation. Market-based schemes, on the other hand, involve creating economic incentives and mechanisms, such as cap-and-trade systems, where tradeable permits are allotted to regulate pollution. Education and awareness programs aim to inform and educate the public about certain issues and encourage them to take action.

In summary, policy instruments and mechanisms play a crucial role in the implementation of policies and the achievement of desired outcomes. Regulations, taxes, subsidies, voluntary agreements, market-based schemes, and education programs are examples of tools that governments and organizations use to shape behavior, protect the environment, improve public health, and promote economic development. By understanding and utilizing these policy instruments and mechanisms effectively, policymakers can design and implement policies that address societal challenges in a comprehensive and sustainable manner.

Chapter 2: Renewable Energy Sources

Renewable energy sources have emerged as an essential aspect of the global transition towards a sustainable and low-carbon future. These sources of energy offer the promise of reducing greenhouse gas emissions, mitigating climate change, and ensuring a secure energy supply for future generations. In this chapter, we will delve into the world of renewable energy, exploring its various types and the potential it holds for meeting the world's energy demands.

1. Solar Energy:

Solar energy is the most abundant renewable energy source on Earth, capable of powering our planet's electricity needs several times over. This energy is harnessed through the use of solar panels or solar thermal collectors, which convert sunlight into usable energy. Photovoltaic cells within these panels capture the sun's radiation and convert it into direct current (DC) electricity, which can be subsequently used or stored in batteries. With advancements in technology and decreasing costs, solar energy has become one of the most rapidly growing sources of renewable energy worldwide.

2. Wind Energy:

Wind energy is another crucial renewable energy source with immense potential. It is derived from harnessing the kinetic energy of the wind through wind turbines. These turbines capture the wind's power and convert it into rotary motion, which is then used to generate electricity. Wind farms, consisting of multiple turbines, have the ability to produce large amounts of energy and can supply power to entire communities or even countries. However, wind energy generation is largely dependent on the availability of strong and consistent winds, which limits its deployment to only suitable areas.

3. Hydropower:

Hydropower, often referred to as hydroelectric power, relies on the energy of moving water to generate electricity. This renewable energy source harnesses the gravitational potential energy of water stored in reservoirs or flowing in rivers and converts it into mechanical energy through the rotating blades of a turbine. The turbine then drives a generator, which produces electricity. Hydropower plants have the advantage of being able to quickly respond to fluctuations in electricity demand, making them a reliable and flexible source of renewable energy.

4. Biomass:

Biomass energy is derived from organic matter, such as plant material and agricultural waste. It can be utilized through various processes, including burning biomass to produce heat, generating biogas from organic waste through anaerobic digestion, or converting biomass into biofuels like ethanol and biodiesel. Biomass energy is considered carbon-neutral because the carbon dioxide released during combustion is reabsorbed by new plants or crops grown for energy production. However, the sustainable sourcing and management of biomass resources are critical to avoid environmental degradation and ensure its long-term viability.

5. Geothermal Energy:

Geothermal energy harnesses the thermal energy stored beneath the Earth's surface. This renewable energy source employs geothermal power plants to extract heat from deep underground reservoirs of hot water or steam. The heat is then used to generate electricity through steam turbines or direct-use applications, such as heating buildings or providing hot water. Geothermal energy is considered reliable and baseload, as it operates continuously with minimal generation fluctuations, thereby contributing to a stable and resilient energy system.

RENEWABLE ENERGY SOURCES offer a diverse and promising array of options for diversifying the world's energy portfolio and transitioning away from fossil fuel dependence. With ongoing technological advancements and decreasing costs, renewable energy has become increasingly competitive. However, challenges remain, including the intermittency of some sources, grid integration, and upfront investment costs. Governments, businesses, and individuals must collaborate to accelerate the adoption of renewable energy, promoting clean and sustainable energy practices to safeguard our planet for future generations.

1. Exploring Renewable Energy Options

When it comes to exploring renewable energy options, there is a vast array of information to consider. Renewable energy is becoming increasingly popular, as people are becoming more aware of the negative impact that fossil fuels have on the environment. In this article, we will delve into the various renewable energy sources, providing details on their benefits and limitations.

One of the most well-known renewable energy sources is solar power. Solar energy is harnessed from the rays of the sun, either through photovoltaic panels or solar thermal systems. One major advantage of solar power is its abundance, as the sun provides an infinite supply of energy. Additionally, solar energy is clean and does not release harmful greenhouse gases. However, a major drawback of solar power is its reliability. Energy production is dependent on weather conditions and the availability of sunlight. Furthermore, the initial cost of installing solar panels can be quite high, making it a long-term investment.

Another important renewable energy source is wind power. Wind turbines generate electricity by harnessing the kinetic energy of the wind. Like solar power, wind power is abundant and does not produce pollutants. Wind farms have the potential to generate massive amounts of electricity and provide a sustainable solution to our energy needs. On the downside, wind energy can be erratic in nature, relying on consistent and appropriate wind speeds for optimal production. Furthermore, wind turbines can be criticized for their visual impact on landscapes and potential negative effects on wildlife.

Hydroelectric power is another major player in the world of renewable energy. This involves capturing the energy from flowing water, such as rivers or dams, and converting it into electricity. Hydroelectric power is reliable and can supply energy continuously, avoiding the intermittency issue faced by solar and wind power. Additionally, hydroelectric plants can also provide other benefits, such as flood control and irrigation. However, the construction of dams and diversion of water can have significant ecological impacts, disrupting aquatic ecosystems and altering natural flow patterns.

Geothermal energy is a lesser-known renewable energy source. It involves utilizing the heat stored within the Earth's crust to generate power. This energy can be accessed through geothermal power plants or directly in the form of heat for residential and commercial use. Geothermal energy has a low environmental impact and can provide a constant and reliable source of power. Nonetheless, the availability of geothermal resources is limited to specific areas with suitable geological conditions.

Biomass energy is an intriguing alternative renewable source. It is generated through the burning of organic materials like wood, crops, and organic waste. This process releases carbon dioxide, but because biomass is a renewable resource and the plants used for biomass naturally absorb carbon dioxide during their lifetime, the net emissions are considered relatively low compared to fossil fuels. Biomass can be converted into biofuels, which can power vehicles. However, the land and resources required for large-scale biomass production raise concerns about food security and sustainability.

In recent years, developments in technology have opened up exciting possibilities for other emerging sources of renewable energy. These include tidal and wave energy, where power is extracted from the movement of ocean currents and waves. Although still in the early stages, the potential for harnessing this vast source of energy is substantial.

It is essential to note that each renewable energy source has its limitations and advantages. The choice of which source to utilize depends on several factors, such as geographic location, financial viability, and environmental impact. Often, a combination of different renewable energy sources is necessary to ensure a reliable and diversified energy supply.

In conclusion, exploring renewable energy options is a critical step towards a sustainable and greener future. Solar, wind, hydroelectric, geothermal, biomass, and emerging technologies all provide viable alternatives to traditional fossil fuel-based energy. By understanding their benefits and limitations, we can make informed decisions when it comes to embracing renewable energy and mitigating environmental damage.

2. Solar Energy: Harnessing the Sun's Power

Solar energy is the harnessing of the sun's power to generate electricity. It is an increasingly popular and environmentally friendly method of generating electricity that has the potential to provide a constant source of clean energy. Unlike traditional fossil fuels, solar energy does not produce harmful greenhouse gas emissions, making it a renewable and sustainable energy source.

One of the main ways solar energy is harnessed is through the use of solar panels, also known as photovoltaic (PV) panels. These panels are made up of numerous individual solar cells that convert sunlight into direct current (DC) electricity. When sunlight hits the solar cells, electrons are freed from their atomic bonds and flow through the material, creating an electric current. This current is then captured by an inverter, which converts it into alternating current (AC), the form of electricity used in homes and businesses.

Solar panels can be installed on rooftops, on the ground, or even integrated into building materials such as windows or walls. The amount of electricity generated by a solar panel depends on several factors, including the size and efficiency of the panel, the intensity of sunlight, and the angle at which the panel is positioned relative to the sun. Generally, larger solar panels with higher efficiency ratings will produce more electricity.

However, solar panels are not the only way to harness the sun's power. Another method is through the use of solar thermal systems, which capture the sun's heat to generate electricity or provide hot water and heating. Solar thermal systems use mirrors or lenses to concentrate sunlight onto a receiver, which absorbs the energy and heats up a fluid like water or oil. This heated fluid can then be used to produce steam, which drives a turbine connected to a generator to produce electricity.

Solar thermal systems can also be used for heating purposes in residential and commercial buildings. In these systems, sunlight is collected and used to heat water or air, which is then circulated throughout the building to provide warmth. Solar thermal systems are particularly effective in sunny regions with high levels of direct sunlight, as they require a considerable amount of sunlight to generate significant amounts of heat.

In recent years, there have been significant advancements in solar energy technology. The efficiency of solar panels has increased, meaning they can generate more electricity from the same amount of sunlight. Additionally, the cost of solar panels has decreased, making them more affordable and accessible to everyday consumers. These advancements, coupled with government incentives and an increased awareness of the need for clean energy, have led to a significant growth in the solar industry.

Despite these advancements, there are still challenges in fully harnessing the sun's power on a large scale. The intermittency of sunlight and the storage of excess electricity continue to be areas of focus for research and development. The development of efficient and cost-effective energy storage solutions, such as batteries, will be crucial in overcoming these challenges and fully realizing the potential of solar energy.

On a global scale, solar energy has the potential to revolutionize the way we generate electricity. It can provide a sustainable, clean, and reliable energy source that can reduce our dependence on fossil fuels and contribute to the fight against climate change. By harnessing the sun's power, we can create a brighter future for generations to come.

3. Wind Energy: Blowing Towards a Sustainable Future

Wind energy, one of the key renewable energy sources, is gaining increasing attention worldwide as a sustainable and environmentally-friendly option for meeting our energy needs. With the many benefits it offers, wind energy is being touted as a key player in the transition towards a clean energy future.

One of the most significant advantages of wind energy is its ability to reduce greenhouse gas emissions. Unlike fossil fuels such as coal, oil, and gas, wind power does not produce harmful carbon dioxide emissions. This makes it a great alternative for combating climate change and reducing our carbon footprint.

Wind energy also offers economic benefits. The installation and maintenance of wind turbines create jobs and stimulate local economies. Furthermore, the production of wind energy eliminates the need for energy imports, reducing dependency on foreign sources and promoting energy independence.

In terms of sustainability, wind energy is unlimited and inexhaustible. Wind turbines use kinetic energy from the wind to generate electricity, and as long as there is wind, there will be power. Unlike fossil fuels, wind energy reserves will never run out, ensuring a long-term and reliable energy source.

In recent years, wind technology has advanced dramatically, making wind energy more efficient and cost-effective. The development of larger turbines, improved manufacturing techniques, and better design have all contributed to increased energy production and lowering costs. Subsidies and government support in many countries have also made wind energy more accessible and financially viable.

Additionally, wind energy has minimal environmental impacts compared to other forms of energy. While some concerns have been raised regarding noise levels, visual aesthetics, and effects on wildlife, the overall impact is far less than that of fossil fuel power plants. Modern wind turbines are built with noise management systems, and careful planning can address any potential conflicts with wildlife.

Wind energy, however, does have some limitations. The most significant is its intermittent nature, as wind is not always constant or predictable. This requires backup power sources or energy storage systems to ensure a stable electricity supply. Nevertheless, advancements in grid technology and energy storage solutions are addressing this challenge. Increased integration with other renewable energy sources, such as solar power, can also help balance the intermittent nature of wind energy.

Furthermore, certain geographical and environmental factors must be taken into account when harnessing wind energy effectively. Wind speeds and patterns vary from region to region, and not all locations are suitable for wind farms. Careful site selection, using data from many years of wind assessment, is necessary to ensure optimal energy production.

In conclusion, wind energy holds immense potential for a sustainable future. Its numerous advantages, such as reduced emissions, economic benefits, and unlimited supply, make it an attractive option for meeting our energy needs. While there are some challenges to overcome, advancements in technology and improved infrastructure are making wind energy increasingly viable. With continued support and investments, wind energy has the potential to play a significant role in our transition towards a cleaner and more sustainable energy future.

4. Hydroelectric Power: The Power of Water

Hydroelectric power, also known as water power, is the energy that is harnessed from the force of moving water. It has been used for centuries as a source of power, and is often viewed as one of the most renewable and sustainable forms of energy available today. In fact, hydroelectric power accounts for approximately 16% of global electricity production, making it one of the most widely used sources of renewable energy worldwide.

The principle behind hydroelectric power is quite simple. It involves the conversion of the potential energy stored in water, such as a river or reservoir, into mechanical energy by using a hydraulic turbine. This mechanical energy is then transformed into electrical energy through the use of a generator. The process may sound complex, but it is actually a very efficient and effective means of generating electricity.

Hydropower plants usually consist of several key components. The first is the dam, which is used to create a reservoir of water. When the gates of the dam are opened, the stored water is allowed to flow through a channel called a penstock. This channel directs the water towards the hydraulic turbine, which is connected to a generator. As the water hits the blades of the turbine, it spins the rotor inside the generator, creating electricity.

One of the advantages of hydroelectric power is its ability to produce large amounts of energy. Dams and reservoirs can store enormous quantities of water, which can be released when needed to generate electricity. Furthermore, the power output of hydropower plants can be easily adjusted, allowing for flexible energy production that can respond to fluctuations in demand.

Another benefit of hydroelectric power is its renewable nature. The water cycle ensures a constant supply of water, making hydropower a reliable and sustainable energy source. Unlike fossil fuels, which require extraction and combustion, hydroelectric power does not emit greenhouse gases or other pollutants during operation. This makes it an environmentally friendly option that helps combat climate change and reduce air pollution.

Hydroelectric power also has several economic advantages. Once the initial infrastructure costs are covered, the operation and maintenance costs of a hydropower plant are relatively low. Additionally, the lifespan of these plants can be quite long, with some facilities still operating after several decades. This longevity allows for continued energy production and a consistent revenue stream.

In addition to its technical and economic benefits, hydropower can also provide social and recreational opportunities. Many reservoirs created by dams serve as attractions for outdoor activities such as boating, fishing, and wildlife observation. The availability of water has fuelled the development of irrigation systems that support agriculture in arid regions, contributing to food security and economic growth.

Despite its numerous advantages, hydroelectric power is not without its drawbacks. One of the major concerns is the ecological impact of dam construction and operation. Dams can disrupt river ecosystems, alter biodiversity, and impact the natural flow of rivers, affecting fish and other wildlife populations. Moreover, large dams can require the displacement of local communities and inundate valuable land.

There is also the issue of sedimentation. Over time, reservoirs accumulate sediment, resulting in reduced storage capacity and decreased power generation efficiency. This requires regular maintenance, such as desilting and dredging, which can be costly and time-consuming.

Despite these challenges, hydroelectric power remains a crucial component of the global energy mix. Its ability to produce large quantities of renewable and sustainable energy allows countries to reduce their dependence on fossil fuels and make significant strides towards a greener future. With further advancements in technology and increased investment in research and development, hydroelectric power has the potential to become an even more reliable, efficient, and environmentally friendly source of energy.

5. Geothermal Energy: Harnessing the Earth's Heat

Geothermal energy is a powerful and sustainable source of energy that harnesses the heat from deep within the Earth. It is a form of renewable energy that has been used for thousands of years, but in recent decades, it has gained significant attention as a viable alternative to fossil fuels.

One of the main advantages of geothermal energy is its abundance. The Earth's core is a constant source of heat, and this energy can be extracted and used to generate electricity and heat. Unlike fossil fuels, which are finite and contribute to greenhouse gas emissions, geothermal energy is essentially unlimited and produces no harmful pollutants.

To tap into geothermal energy, wells are drilled deep into the Earth's crust to access high-temperature reservoirs of steam or hot water. The steam or hot water is then brought to the surface and used to power turbines, which generate electricity. In some cases, direct use systems are employed, where the hot water is used directly for heating purposes.

Geothermal power plants can be divided into three main types: dry steam, flash steam, and binary cycle. Dry steam power plants utilize steam directly from the geothermal reservoir to generate power. Flash steam plants, on the other hand, use hot water from the reservoir, which is then flashed to steam by reducing its pressure. Binary cycle power plants are the most common type and use lower-temperature geothermal resources by utilizing a secondary fluid with a lower boiling point to drive the turbines.

One of the major limiting factors of geothermal energy is the location of geothermal reservoirs. They are typically found in areas where the Earth's crust is thin enough to allow for drilling, such as volcanic regions or areas with high tectonic activity. Nonetheless, advancements in technology have allowed for deeper drilling and the utilization of lower-temperature resources, broadening the potential geothermal sites.

Geothermal energy has several benefits that make it an attractive option for energy production. Firstly, it is highly reliable. Unlike solar or wind energy, geothermal power plants can operate continuously and produce a consistent supply of electricity, regardless of weather conditions. This makes it a valuable baseload energy source, which complements the intermittent nature of renewable energy sources.

Furthermore, geothermal energy has a small physical footprint compared to other forms of energy production. Power plants can be built underground, reducing visual impact and land disturbance. Additionally, modern geothermal power plants have a relatively low environmental impact, as they produce zero greenhouse gas emissions and minimal noise pollution.

Geothermal energy also has the advantage of being a local resource, reducing dependence on imported energy sources. Rural communities and remote areas that are not connected to the grid can benefit greatly from geothermal energy, as it can provide them with a sustainable and reliable source of electricity and heat.

In conclusion, geothermal energy is a fascinating and promising renewable energy source that harnesses the Earth's heat. With advancements in technology, the potential for geothermal power has expanded, allowing for the utilization of lower-temperature resources and the development of power plants in various locations. Its reliability, environmental benefits, and local availability make it an attractive option for sustainable energy production. As the world continues to seek alternative energy sources, geothermal energy has the potential to play a significant role in the transition to a more sustainable future.

6. Biomass Energy: Utilizing Organic Resources

Biomass energy is a type of renewable energy that involves the utilization of organic materials- such as plants, agricultural crops, animal waste, and forestry byproducts- to generate electricity or produce heat. This form of energy is considered renewable because organic materials can be replenished through natural processes over time.

One key advantage of biomass energy is its ability to provide a sustainable, carbon-neutral alternative to fossil fuels. When organic materials are burned for energy, they release carbon dioxide- a greenhouse gas- into the atmosphere. However, since the plants and trees used in biomass energy production absorb carbon dioxide during their growth, the overall carbon emissions are balanced out. In this way, biomass energy can play a crucial role in mitigating climate change and reducing the reliance on non-renewable energy sources.

There are various methods of harnessing biomass energy, each with its unique advantages and applications. One common approach is the direct combustion of biomass materials in specialized power plants. In these facilities, organic resources are burned to heat water and produce steam, which then drives a turbine to generate electrical power. This method is widely used across the globe and is relatively simple and cost-effective to implement, especially when existing power plants are converted from fossil fuels to biomass.

Another prominent method of utilizing biomass energy is through the production of biogas. This involves harnessing the methane gas produced from the breakdown of organic materials such as agricultural waste, manure, and food waste in anaerobic digestion facilities. Biogas can be used directly for heating or converted into electricity and heat using biogas-fired power plants. Additionally, biogas can be upgraded to natural gas quality and injected into existing gas grids, allowing for its distribution and use in a variety of sectors, including transportation.

In recent years, advancements in technology have allowed for the production of biofuels from biomass sources. These fuels, which include ethanol and biodiesel, can be utilized as a direct replacement for conventional gasoline and diesel fuels in transportation. Ethanol, for example, can be obtained through the fermentation of sugar-rich crops such as corn, sugarcane, and wheat. Biodiesel, on the other hand, can be produced by reacting plant or animal oils with an alcohol.

The production of biomass energy has numerous environmental benefits apart from its carbon neutrality. Firstly, utilizing organic waste for energy can reduce landfill quantities and the associated emission of potent greenhouse gases such as methane. Additionally, the cultivation of dedicated energy crops can lead to improvements in soil quality, increased biodiversity, and the reduction of soil erosion, providing additional environmental benefits.

However, despite its advantages, biomass energy also faces certain challenges and controversies. The cultivation of energy crops, for instance, may compete with food production and result in deforestation or unsustainable land-use practices. It is essential to approach biomass energy production with careful planning and ensure that the resources used for energy production do not compromise food security or the natural balance of ecosystems.

Nevertheless, the utilization of organic resources for biomass energy has significant potential for sustainable energy generation. With proper management and responsible practices, biomass energy can contribute to the transition towards a cleaner and more secure energy future. As technology continues to advance, the role of biomass energy is likely to expand further, offering innovative solutions to tackle climate change and promote a greener society.

Chapter 3: Sustainable Agriculture and Food Systems

In Chapter 3 of our comprehensive study on sustainable agriculture and food systems, we delve deep into the intricate world of sustainable agriculture practices and the potential benefits they hold for our future. Covering a myriad of topics, the chapter presents a detailed analysis of sustainable agriculture principles, methodologies, and innovations that are transforming the food production landscape.

One of the main focuses of this chapter is the concept of agroecology – a holistic approach to agricultural production that takes into account the ecological, social, and economic interactions of the entire farming system. We discuss the importance of diversifying agricultural landscapes and promoting biodiversity, as well as the significance of crop rotation, cover cropping, and intercropping techniques in maintaining soil health and fertility while minimizing the need for synthetic inputs.

The chapter unearths the potential of organic farming, showcasing its numerous advantages over conventional practices. We explore the environmental benefits of organic farming, such as reduced water pollution, carbon sequestration, and preservation of ecosystems. We also shed light on the health benefits of consuming organic produce, as they are free from synthetic pesticides and genetically modified organisms (GMOs), making them a healthier choice for both consumers and farmers.

Moreover, we venture into the realm of precision agriculture and its role in achieving sustainable agriculture. Through advanced technologies, such as remote sensing, geographic information systems (GIS), and the Internet of Things (IoT), precision agriculture aims to optimize resource utilization and minimize waste. We delve into the impact of precision agriculture on reducing greenhouse gas emissions, increasing water use efficiency, and reducing nutrient runoff, all contributing to more sustainable farming practices.

Furthermore, we explore the vital connections between sustainable agriculture and food security. By adopting sustainable techniques, farmers can improve the productivity of their land in a manner that is sustainable in the long-term, ensuring reliable food production for a growing global population. Additionally, sustainable agriculture practices often prioritize local and regional markets, promoting food sovereignty and reducing dependency on international food trade.

The chapter also highlights successful case studies from around the world, where sustainable agriculture projects have transformed communities and demonstrated the viability of sustainable food production systems. These stories serve as inspiration and model the potential for sustainable practices to address food security, environmental conservation, and social justice challenges simultaneously.

Without a doubt, Chapter 3 of our study on sustainable agriculture and food systems is an invaluable resource for anyone interested in better understanding the intricacies of sustainable farming practices and the transformative effects they can have on our food systems. From agroecology to precision agriculture, and from organic farming to food security, this chapter provides a comprehensive, informative, and thought-provoking analysis that will lead readers towards a deeper understanding of sustainable agriculture's potential.

1. Transitioning to Sustainable Agriculture

Transitioning to sustainable agriculture is becoming more imperative as the negative environmental effects of conventional farming practices become increasingly evident. Sustainable agriculture aims to provide long-term ecological balance, economic viability, and social equity by implementing practices that preserve soil health, conserve water, promote biodiversity, and minimize the use of synthetic inputs.

One of the key aspects of transitioning to sustainable agriculture is adopting organic farming practices. Organic farming eliminates the use of synthetic pesticides, herbicides, fertilizers, and genetically modified organisms (GMOs). Instead, it relies on natural methods such as crop rotation, organic fertilizers, and biological pest control. This shift to organic practices not only reduces the use of harmful chemicals that damage the environment and endanger human health but also helps to maintain soil fertility and improve the nutritional quality of crops.

Another important element of sustainable agriculture is promoting soil health. Well-managed soils are crucial for sustainable farming as they support plant growth, retain water, and increase nutrient availability. Practices such as cover cropping, crop rotation, and the use of organic matter like compost help to build soil organic matter and enrich soil fertility. By improving soil health, farmers can minimize erosion, enhance soil structure, and increase the water-holding capacity of the soil, ultimately leading to increased productivity and resilience of agricultural systems.

Conserving water is another major concern in sustainable agriculture. With increasing water scarcity and climate change, efficient water management is vital for agricultural sustainability. Practices such as drip irrigation, rainwater harvesting, and moisture sensors can help optimize water usage, reduce water wastage through evaporation or runoff, and ensure that water is used effectively without compromising crop yields. Additionally, improving soil health through organic farming practices also enhances the soil's ability to retain water, further contributing to water conservation efforts.

Biodiversity conservation is another crucial aspect of sustainable agriculture. Since agriculture often leads to habitat destruction, monocultures, and the use of chemical inputs that harm beneficial species, promoting biodiversity is essential for the long-term viability of agricultural ecosystems. Planting native species, creating wildlife habitats, and implementing agroforestry systems can help support a diverse range of organisms, including beneficial insects, birds, and pollinators, which in turn contribute to effective pest control and pollination.

Furthermore, the reduction of synthetic inputs, such as chemical fertilizers and pesticides, is a pivotal part of sustainable agriculture. Excessive use of these inputs can lead to pollution of soil, water bodies, and air, negatively impacting human health and ecosystems. By embracing various innovative methods, such as integrated pest management and precision agriculture, farmers can minimize their reliance on synthetic inputs. Integrated pest management methods favor biological control agents to combat pests, while precision agriculture employs technology to target inputs and optimize their usage based on specific crop requirements, reducing overall chemical usage.

Overall, transitioning to sustainable agriculture requires a shift in mindset and practices on a global scale. Governments, farmers, researchers, and consumers all play a critical role in promoting and implementing sustainable agricultural practices. Governments should provide support through incentives, policics, and regulations that encourage sustainable farming methods. Farmers need access to training, resources, and financial aid to transition from conventional to sustainable practices. Researchers must continue to develop and disseminate knowledge about sustainable agricultural approaches, while consumers can contribute by supporting local and organic products and being mindful of their food choices.

In conclusion, transitioning to sustainable agriculture is necessary to ensure food security, protect the environment, and fulfill our moral obligation to future generations. By adopting organic farming practices, promoting soil health, conserving water, conserving biodiversity, and reducing synthetic inputs, we can mitigate the negative impacts of conventional farming and pave the way for a more sustainable and resilient agricultural system. The transformation is complex and requires a collaborative effort, but the rewards of sustainable agriculture will benefit not only the present but also the future.

2. Organic Farming Practices

Organic farming practices have gained popularity in recent years due to their sustainable and environmentally friendly approach towards agriculture. Unlike conventional farming, which heavily relies on synthetic chemicals, organic farming emphasizes natural farming methods that promote soil health, biodiversity, and overall sustainable food production.

One of the key principles of organic farming is the prohibition of synthetic pesticides, herbicides, and fertilizers. Instead, organic farmers adopt various techniques to maintain soil fertility and control pests and diseases without the use of harmful chemicals. For instance, they focus on building healthy soil by using compost, animal manure, and cover crops. These practices not only enrich the soil with essential nutrients but also improve its structure and ability to retain water.

Crop rotation is another important technique used in organic farming. It involves growing different crops in a specific sequence on a specific piece of land. This practice helps improve soil fertility by reducing the risk of nutrient depletion and pest infestation. Furthermore, incorporating legume crops into the rotation helps fix atmospheric nitrogen into the soil, leading to enhanced nitrogen availability for subsequent plantings.

Weed management in organic farming is primarily done through cultural and mechanical methods. Organic farmers employ techniques like hand weeding, mulching, crop rotation, and the use of physical barriers to control the growth and spread of weeds. Although these methods may require more labor compared to conventional herbicide use, they promote a healthier farming environment while minimizing potential chemical residues in crops.

To control pests and diseases, organic farmers rely on biological control methods. These methods involve promoting the presence of beneficial organisms such as predatory insects, birds, and certain microorganisms that prey on pests. Additionally, farmers may use physical barriers, traps, and pheromone-based pest control techniques to manage pests effectively. By mimicking nature's balance, organic farming reduces the dependence on chemical pesticides, thus reducing harm to beneficial insects, birds, and other wildlife.

Another essential aspect of organic farming is livestock management. Organic livestock must be raised in conditions that prioritize their welfare. Animals must have access to pasture and be fed organic feed without the use of growth hormones or antibiotics. This ensures the production of organic meat, dairy, and other animal products that are free from residue traces of synthetic chemicals.

Organic farming practices go beyond just avoiding the use of synthetic chemicals. They aim to create a harmonious system that works in synchronization with nature's cycles. By emphasizing soil health and biodiversity, organic farming contributes to the conservation of ecosystems and the promotion of sustainable agriculture.

From an environmental perspective, organic farming practices can significantly reduce pollution and resource consumption. Organic farms tend to have lower energy usage, as they do not rely on synthetic fertilizers or large-scale processing. Water conservation is also emphasized, as organic farmers prioritize efficient irrigation techniques and soil management practices that minimize water loss through runoff.

In addition to environmental benefits, there are numerous advantages of organic farming from a societal standpoint as well. Organic farms often support local economies by selling their produce within a smaller radius, decreasing transportation distances, and thereby reducing overall carbon emissions. In many cases, organic farming is also labor-intensive, creating job opportunities in rural areas and contributing to sustainable livelihoods.

Consumers are increasingly drawn to organic produce due to its perceived health benefits. Organic crops are believed to contain higher nutrient levels and lower pesticide residue compared to conventionally produced crops. Though scientific evidence remains inconclusive regarding these claims, demand for organic products continues to rise, reflecting the growing interest in sustainable and healthier food choices.

While organic farming has its advantages, it also faces certain challenges. Transitioning from conventional to organic farming requires significant time, effort, and risk for farmers. The initial adjustment period may lead to lower yields and higher production costs. However, in the long run, the investment in organic farming practices can pay off through improved soil fertility, reduced input costs, and the potential for higher market prices.

In conclusion, organic farming practices offer a sustainable and eco-friendly alternative to conventional agriculture. Through techniques such as natural pest control, crop rotation, and the enrichment of soil health, organic farming promotes biodiversity, conserves resources, and produces healthier food. While challenges may arise during the transition, the long-term benefits of organic farming justify the efforts to create a more sustainable future for agriculture and food production.

3. Locally Sourced Food and Urban Farming

Locally sourced food and urban farming have become increasingly popular topics in recent years, as more and more people are concerned about where their food comes from and the environmental impact of industrial agriculture. In this article, we will explore the benefits and challenges of locally sourced food and urban farming, as well as the innovative solutions that are emerging to address these issues.

Locally sourced food refers to food that is grown or produced within a certain radius of where it is consumed. This could be a few miles or a few hundred miles, depending on the definition used. The main advantage of locally sourced food is that it reduces the distance between the farm and the consumer, which means less transportation, refrigeration, and packaging are required. As a result, locally sourced food has a lower carbon footprint compared to food that travels long distances.

Another benefit of locally sourced food is that it supports the local economy. By purchasing food from local farmers, consumers are directly contributing to the livelihoods of these farmers and the overall economic well-being of their community. Additionally, locally sourced food tends to be fresher, as it is harvested at its peak and does not need to endure long journeys or extensive periods of cold storage. This means it is often more flavorful and nutrient-rich.

Urban farming, on the other hand, involves the cultivation of food within cities or urban areas. It can take many forms, including rooftop gardens, vertical farming, and community gardens. Urban farming allows individuals and communities to grow their own food, promoting self-sufficiency and providing access to fresh produce in areas where it may be limited.

One of the major advantages of urban farming is that it reduces the environmental impact of food production. By growing food in cities, we can significantly reduce the distance that food needs to travel to reach consumers.

This not only reduces carbon emissions but also decreases the need for land and water resources, which are scarce in urban areas.

Furthermore, urban farming can help address food security issues, particularly in low-income neighborhoods where access to fresh, nutritious food may be limited. Community gardens and urban farms provide opportunities for individuals to grow their own food, improving food access, and increasing food sovereignty. They also create spaces for education and community engagement, promoting sustainable practices and healthy eating habits.

However, there are several challenges associated with locally sourced food and urban farming. Firstly, land availability can be a significant limitation in urban areas. As cities become denser, open spaces suitable for farming become scarce. This obstacle can be overcome through the use of innovative farming techniques such as hydroponics and vertical farming, which require less space and can be implemented in small or unconventional areas.

Additionally, urban farming may face challenges related to soil contamination. Urban areas are often polluted with heavy metals and other toxins, which can make the soil unfit for farming. This issue can be addressed by using raised beds or containers and properly testing and treating the soil to ensure it is safe for growing food.

Furthermore, there may be regulatory and zoning restrictions that prohibit or limit urban farming activities. To promote and support urban agriculture, policymakers need to create regulations that enable and encourage its development. This could include establishing urban farming zones, providing financial incentives or grants for urban farmers, or revising building codes to accommodate rooftop and vertical farming.

Despite these challenges, individuals, communities, and organizations are actively finding solutions and driving the growth of locally sourced food and urban farming. Initiatives such as farmers' markets, community-supported agriculture (CSA) programs, and food cooperatives connect consumers with local farmers and ensure a steady market for their produce. Innovative initiatives like farm-to-table restaurants and urban farming cooperatives are redefining the way food is produced and consumed, making it more sustainable and supportive of local economies.

In conclusion, locally sourced food and urban farming offer numerous benefits for individuals, communities, and the environment. They reduce carbon emissions, support local economies, and improve food access and security. However, there are challenges to overcome, such as the availability of land and soil contamination. Nevertheless, with the growing interest and determination to create a sustainable and resilient food system, the future of locally sourced food and urban farming looks promising.

4. Innovations in Food Distribution and Storage

Innovations in food distribution and storage have played a crucial role in enhancing the efficiency, quality, and sustainability of the food supply chain. As the global population continues to grow and urbanize, there is an increasing demand for safe and accessible food that can be distributed quickly and stored for extended periods of time. This has spurred a wave of technological advancements and novel ideas that have revolutionized the way food is distributed and stored.

One of the most significant innovations in food distribution is the advent of refrigerated transportation. Before the invention of refrigerated trucks and containers, perishable foods like meat, dairy products, and fresh produce were largely limited to local markets. However, with the ability to maintain a cool environment during transit, it became possible to transport perishable goods across long distances. This not only expanded the variety of foods available to consumers but also opened up new markets for farmers and producers. In addition, refrigeration has played a crucial role in reducing food waste by prolonging the shelf life of perishable items.

Another notable development in food distribution is the rise of e-commerce. The advent of the internet and online platforms has greatly enhanced the accessibility and convenience of food delivery. Companies like Amazon, FreshDirect, and Ocado have capitalized on this trend by developing sophisticated logistics networks and cold chain infrastructure for efficient delivery of perishable goods. Consumers can now order groceries online with just a few clicks and have them delivered straight to their doorstep. This not only saves time and effort but also allows for better inventory management and reduced food waste. Additionally, smart technology-enabled delivery options, such as refrigerated lockers and drones, are being explored to further optimize the last-mile distribution.

Innovative packaging materials and techniques have also had a significant impact on food storage. Advances in materials science have led to the development of antimicrobial packaging, which helps extend the shelf life of products and prevents spoilage. These materials release active compounds that kill or inhibit the growth of bacteria and fungi, thereby keeping food fresh for longer periods. Furthermore, vacuum packaging, modified atmosphere packaging (MAP), and controlled atmosphere packaging (CAP) have gained popularity for their ability to maintain and control the environment within a package, reducing the deterioration of food.

Moreover, smart technologies and data analytics have been increasingly employed to monitor and manage food supply chains. Internet-of-Things (IoT) devices and sensors enable real-time tracking of temperature, humidity, and other environmental parameters during storage and transportation. This allows for early detection of any deviations that could compromise food safety or quality. Advanced analytics platforms process this data to provide valuable insights and optimize the supply chain, reducing inefficiencies and enhancing traceability.

Furthermore, the recent emergence of blockchain technology has also shown promise in improving food distribution and storage. Blockchain provides a tamper-proof and transparent system for recording and verifying transactions, making it ideal for establishing trust and traceability in complex supply chains. With blockchain, retailers and consumers can easily access information about the origin, handling, and conditions of the food they purchase, ensuring its safety and quality.

In conclusion, innovations in food distribution and storage have had a profound impact on the efficiency, safety, and sustainability of the food supply chain. Refrigerated transportation, e-commerce, advanced packaging materials, smart technologies, and blockchain have all contributed to ensuring that food reaches consumers in a timely and secure manner. These advancements not only enhance the accessibility and quality of food but also help reduce food waste and improve transparency. As technology continues to evolve, we can expect further innovations that will shape the future of food distribution and storage.

Chapter 4: Waste Management and Circular Economy

In Chapter 4 of our book, we delve into the world of waste management and the concept of a circular economy. This topic may not immediately capture your attention, but I promise you, it is a profound one that affects every aspect of our lives. From the food we eat to the products we consume, waste management plays a vital role in preserving our environment and ensuring a sustainable future for generations to come.

We start by examining the current state of waste management globally. It is startling to learn that over 2 billion tons of waste is generated each year across the globe. This staggering amount has significant implications for climate change, pollution, and resource scarcity. If we are to mitigate these issues, we must introduce strategies and technologies that allow us to break free from our linear "take-make-dispose" economic model.

And so, we introduce the concept of a circular economy. Unlike our current linear system, which follows a straight line from production to disposal, the circular economy aims to close the loop by maximizing the value of products and materials, and minimizing waste and resource consumption. This shift in mindset requires a change at every level of society- from policymakers and manufacturers to consumers and individuals.

Within this circular economy framework, waste management takes on a completely different perspective. It ceases to be solely about the collection and disposal of waste, but rather becomes a means of extracting value and resources from discarded materials. This can be achieved through various approaches such as recycling, upcycling, and reusing.

One fascinating aspect of waste management within the circular economy is the concept of industrial symbiosis. This refers to the collaboration between different industries to exchange waste and by-products, thus creating a closed-loop system of resource utilization. For example, a brewery may provide its spent grain to a local farm for animal feed, while receiving in return organic fertilizers or even biogas for energy generation. These symbiotic relationships not only reduce waste and pollution but also create economic and environmental benefits for all involved parties.

Furthermore, advancements in technology play a crucial role in waste management and the circular economy. From waste-to-energy technologies that transform waste into heat and electricity, to material recovery facilities that sort and process recyclable materials efficiently, there are numerous innovative solutions to combat the waste crisis. These technologies not only assist in waste management but also present new economic opportunities and job creation potential.

Additionally, waste management must involve behavioral changes and public education. Although advancements in technology are crucial, it is equally important to recognize the importance of individual actions. This includes proper waste sorting, conscientious consumer choices, and reducing reliance on single-use items. By championing sustainable lifestyles and promoting environmental awareness, we can collectively contribute to an effective waste management system.

In conclusion, waste management and the circular economy are more than just buzzwords. They represent a paradigm shift in how we interact with our environment, consume resources, and build our societies. By adopting a circular mindset, we can transform waste from a problem into an opportunity, where what was once discarded becomes a valuable resource for future generations. It is an ambitious goal, but one that we simply cannot afford to ignore.

1. The Challenges of Waste Management

Waste management is a crucial issue that is facing societies all over the world. With the increase in population and consumption levels, the amount of waste being generated is also multiplying at an alarming rate. This poses significant challenges for waste management systems to effectively deal with the growing problem.

One of the main challenges of waste management is the sheer quantity of waste being generated. Urban areas, in particular, produce enormous amounts of waste on a daily basis. From household garbage to commercial and industrial waste, the volume is overwhelming. This presents a logistical challenge in terms of adequately collecting and disposing of all this waste in a timely and efficient manner.

Another challenge is the diversity of waste types. Waste comes in various forms, from solid to liquid, hazardous to non-hazardous. Different waste materials require different treatment and disposal methods. For instance, medical waste needs to be treated and disposed of carefully to prevent any environmental or health hazards. Chemical waste must be managed in a way that minimizes risks to human health and the environment. Dealing with this diverse array of waste materials requires specialized infrastructure, equipment, and expertise.

Furthermore, waste management poses financial obstacles. Implementing effective waste management systems requires significant investments in infrastructure, technology, and human resources. Municipalities and governments may struggle to secure the necessary funding to establish and maintain a comprehensive waste management system. Moreover, waste management costs often fall on the local authorities and taxpayers, leading to debates and disputes over who bears the financial burden. Balancing environmental concerns with financial constraints is a complex and challenging task for policymakers.

In addition, waste management also faces the challenge of changing societal attitudes and behaviors. In many countries, dumping waste irresponsibly or burning it openly is still prevalent due to a lack of awareness or disregard for environmentally friendly practices. Public education and awareness campaigns are necessary to promote waste reduction and recycling to mitigate the impact of waste on the environment. However, changing human behavior is a slow and challenging process that requires sustained efforts and effective communication strategies.

Environmental sustainability is another major challenge of waste management. Simply disposing of waste in landfills is not a viable solution in the long run. Landfills take up valuable land space, can contaminate groundwater, and emit harmful gases, contributing to air pollution and climate change. Therefore, waste management systems need to focus on waste reduction, reuse, and recycling to minimize the amount of waste ending up in landfills. Developing effective recycling programs, promoting circular economies, and investing in innovative waste-to-energy technologies are essential steps towards achieving environmental sustainability in waste management.

Lastly, the globalization of waste is an emerging challenge that waste management must contend with. With global trade and mobility, waste can be shipped across borders or illegally dumped in developing countries. This not only exacerbates the waste management problem, but it also raises ethical concerns related to environmental justice. Collaborative efforts between nations, stricter regulations, and enforcement of international waste disposal standards are necessary to tackle this global challenge.

In conclusion, waste management poses a series of multifaceted challenges. Dealing with the increasing quantity and diversity of waste requires comprehensive infrastructure, equipment, and expertise. Addressing the financial burdens associated with waste management while promoting sustainability is a delicate balancing act. Changing societal attitudes towards waste disposal practices and mitigating the globalization of waste are also significant challenges that require concerted efforts. However, these challenges can be overcome through innovative solutions, effective policies, and global cooperation to create a more sustainable future.

2. Strategies for Effective Waste Reduction

Effective waste reduction strategies play a crucial role in minimizing the environmental impact of our daily activities, conserving natural resources, and promoting sustainability. Here, we will explore two strategies that are known for their effectiveness in waste reduction: recycling and composting.

1. Recycling:

Recycling is a process of converting waste materials into new products, thereby reducing the consumption of raw materials and energy. It involves collecting, sorting, processing, and transforming waste materials into useful materials for manufacturing new products. Here are some key points for effective recycling:

a) Waste segregation and collection: Proper waste segregation is essential to enable efficient recycling. Designated bins for different types of recyclable materials, such as paper, plastic, glass, and metal, should be placed in public places, offices, and households. Encouraging people to separate their waste correctly is crucial to maximize the recycling rate.

b) Education and awareness: Regular awareness campaigns and educational programs can help create a culture of recycling. Educating people about the importance of recycling, the types of materials that can be recycled, and the process involved can significantly boost participation and compliance.

c) Expanded recycling infrastructure: To further enhance recycling efforts, it is important to establish and expand recycling infrastructure. This includes building recycling plants and facilities, promoting recycling centers or drop-off points, and providing support for companies engaging in recycling activities.

d) Incorporating recycled materials: Encouraging industries to use recycled materials in their manufacturing processes should be a priority. Government policies and incentives can promote the use of recycled materials, further boosting the demand for recycling and ensuring that the loop is closed.

2. Composting:

Composting is a process that turns organic waste, such as food scraps and yard trimmings, into nutrient-rich soil. By composting, we divert organic waste from landfills, reduce methane emissions (a potent greenhouse gas), and create nutrient-dense compost for use in gardening, agriculture, and landscaping. Here's how to implement effective composting strategies:

a) Education and promotion: Similar to recycling, raising awareness about the benefits of composting and providing guidance on how to compost correctly is essential. Community workshops, school programs, and public campaigns can educate individuals on the fundamentals of composting and its positive environmental impact.

b) Backyard and community composting: Encouraging individuals to compost at home with small compost piles or utilizing vermicomposting (using worms) can help reduce organic waste at the household level. Similarly, communities can establish centralized composting facilities, enabling residents to contribute their organic waste.

c) Composting at commercial and industrial levels: It is equally important to implement composting practices in larger-scale operations, such as restaurants, hotels, and food processing facilities. Encouraging these establishments to separate organic waste and utilize composting facilities can significantly reduce their waste output.

d) Municipal-level composting: Municipal composting facilities can handle a larger volume of organic waste. Municipalities can partner with private companies or allocate resources to establish comprehensive organic waste management solutions, ensuring a high recycling rate that benefits the community and environment.

Implementing these strategies requires collective effort, commitment, and investment from governments, businesses, and individuals. However, the long-term benefits of effective waste reduction, such as improved environmental conditions, resource conservation, and reduced greenhouse gas emissions, make these efforts worthwhile. By integrating recycling and composting practices into our daily lives, ordinary individuals can become champions of waste reduction and contribute to building a cleaner and more sustainable future.

3. Recycling: Closing the Loop

Recycling is a vital process that plays a crucial role in reducing waste, conserving resources, and minimizing the negative impacts of production and consumption on the environment. By closing the loop in the consumption cycle, recycling ensures that materials are reused and kept in circulation rather than being disposed of in landfills or incinerated.

Closing the loop refers to the idea that, through recycling, the materials used in products can be recovered and transformed into new products. This process not only saves resources but also helps to reduce energy consumption and greenhouse gas emissions associated with the production of new materials. It enables the sustained use of valuable resources, breaks the linear pattern of extraction-production-disposal, and creates a more sustainable and circular economy.

One key aspect of closing the loop is the efficient collection and sorting of recyclable materials. Recycling programs need to be implemented worldwide, and citizens should be educated and encouraged to participate. Efficient recycling systems require convenient and easy-to-use collection bins, as well as clear labeling and guidelines on what can and cannot be recycled. Furthermore, public awareness campaigns are essential for promoting recycling and increasing participation rates.

Another important factor in closing the loop is the development and advancement of recycling technologies. New and innovative processes need to be developed to enable the effective transformation of waste materials into high-quality reusable resources. For example, advancements in plastic recycling technologies have allowed for the production of high-grade plastic pellets, which can be used to create new products. Similarly, developments in e-waste recycling have made it possible to recover valuable metals and components from discarded electronic devices.

In addition to technological advancements, it is crucial to establish market demand for recycled materials. Without a steady demand for recyclables, the entire recycling process can be rendered ineffective. Governments, businesses, and consumers must collectively support the use of recycled materials by purchasing recycled products. By creating demand for recycled materials, markets can be stimulated, investment in recycling technologies can be encouraged, and the loop can be effectively closed.

Lastly, closing the loop also involves encouraging a shift towards a more sustainable and circular design mindset. Products should be designed with recyclability in mind, ensuring that they can be easily disassembled and their components can be recovered for recycling. Designers need to consider the entire life cycle of their products, from the sourcing of raw materials to the eventual disposal or recycling. By incorporating closed-loop design principles, manufacturers can create products that minimize waste generation and maximize recyclability.

In conclusion, closing the loop in recycling is an essential step towards building a more sustainable and circular economy. It requires the implementation of efficient recycling systems, the development of advanced recycling technologies, the creation of market demand for recycled materials, and a shift towards closed-loop design principles. By actively participating in and supporting recycling efforts, individuals, businesses, and governments can collectively contribute to the creation of a better and more environmentally conscious future.

4. Creating a Circular Economy: From Linear to Circular

Creating a Circular Economy: From Linear to Circular

The concept of a circular economy has gained significant attention in recent years, with businesses and governments increasingly recognizing the need to move away from the traditional linear economic model. In a linear economy, resources are extracted, processed into products, consumed, and then discarded as waste. This model is unsustainable, leading to a host of environmental and social challenges.

A circular economy, on the other hand, aims to eliminate waste and maximize the use of resources by keeping materials in use for as long as possible. It operates on the principles of reducing, reusing, repairing, and recycling, with the ultimate goal of closing the loop and creating a regenerative system. This transformative approach requires fundamental changes in the way products are designed, produced, and consumed.

One of the key strategies in transitioning from a linear to a circular economy is adopting a cradle-to-cradle approach. This concept, pioneered by architect William McDonough and chemist Michael Braungart, emphasizes the idea that products should be designed with the intention of being reused or recycled at the end of their useful life. This means considering the entire lifecycle of a product, from the sourcing of materials to their eventual disposal.

Designing products for durability and repairability is a fundamental step in creating a circular economy. By ensuring that products are built to last, and that components can be easily replaced or fixed, we can extend their lifespan and reduce the need for constant replacement. This approach also encourages the development of local repair and refurbishment businesses, which not only create job opportunities but also promote a culture of repair rather than disposal.

Another important aspect of transitioning to a circular economy is shifting from a linear supply chain to a closed-loop system. This involves rethinking our production and consumption patterns to prioritize materials that can be recycled or composted. It also requires implementing effective waste management systems and promoting the use of recycled materials in the manufacturing process. Adopting innovative technologies such as 3D printing can further facilitate the closed-loop system by allowing for the creation of products on-demand, reducing the need for large-scale production and transportation.

In addition to these measures, incentivizing the adoption of circular practices is crucial for driving the transition. Governments can play a significant role in this regard, by providing financial or regulatory incentives for businesses to embrace circularity. These incentives may include tax breaks for companies that use recycled materials, grants for research and development of circular technologies, or stricter regulations on waste management and disposal.

Education and awareness are also key in fostering a transition to a circular economy. By promoting the importance of sustainable consumption and the benefits of circular practices, we can encourage individuals and communities to embrace this transformative approach. This could involve educational campaigns, public awareness initiatives, and collaboration with educational institutions to incorporate circular economy principles into their curriculum.

Ultimately, creating a circular economy requires a comprehensive and collaborative effort from all stakeholders – governments, businesses, consumers, and civil society organizations. It requires a shift in mindset, challenging the prevailing notion that growth and resource consumption are inherently linked. By embracing circularity, we have the opportunity to move towards a more sustainable and resilient economy that respects planetary boundaries and promotes social well-being.

Chapter 5: Eco-friendly Transportation

Transportation is a critical component of our daily lives, enabling us to commute, travel, and transport goods over short and long distances. Unfortunately, traditional modes of transportation have heavily relied on fossil fuels, leading to soaring greenhouse gas emissions and environmental degradation. However, with the increasing urgency to address climate change, the need for eco-friendly transportation options has become evident. This chapter delves into the realm of eco-friendly transportation, providing a comprehensive overview of various sustainable alternatives and their impact on the environment.

1. Electric Vehicles (EVs):

The emergence of electric vehicles marks a significant turning point in the transportation industry's environmental footprint. EVs run on electricity and produce zero tailpipe emissions, resulting in reduced air pollution and improved air quality. Furthermore, as the energy grid becomes greener, the overall carbon footprint of EVs decreases exponentially. With the advancements in battery technology and the development of a widespread charging infrastructure, electric vehicles have become a viable and sustainable mode of transportation.

2. Public Transportation:

Combating the negative environmental consequences of individual car usage requires a shift towards public transportation systems. Buses, trains, trams, and subways offer a more sustainable alternative, as they can transport a larger number of people using significantly less fuel per person. Additionally, the incorporation of renewable energy sources to power public transportation further enhances its green credentials.

3. Cycling and Walking:

One of the simplest, yet often overlooked, forms of eco-friendly transportation is cycling and walking. These modes not only contribute to reducing carbon emissions but also promote physical health, reducing the burden on healthcare systems. Encouraging urban planning that prioritizes pedestrian and cyclist infrastructure becomes imperative to facilitate and encourage active modes of transportation.

4. Carpooling and Ride-Sharing:

Utilizing ridesharing platforms and carpooling arrangements offers an effective way to reduce the number of vehicles on the road, leading to decreased fuel consumption and emissions. In addition to environmental benefits, carpooling and ride-sharing initiatives also provide economic advantages by reducing personal transportation costs and alleviating traffic congestion in urban areas.

5. Sustainable Aviation:

Aviation is a sector notorious for its large carbon footprint. However, efforts are being made to develop sustainable aviation fuels and explore electric or hydrogen-powered aircraft. Additionally, advancements in aircraft design, such as lighter materials and more efficient engines, contribute to reducing the emissions associated with flying, making air travel a more sustainable mode of transportation.

6. Alternative Fuels:

Beyond electric vehicles, several alternative fuels show promise in the quest for eco-friendly transportation. These include biofuels derived from organic materials, such as algae or agricultural waste, which can power vehicles and minimize greenhouse gas emissions. Hydrogen fuel cell technology, which produces water vapor as the lone emission, also presents a sustainable option for powering various transportation modes.

ECO-FRIENDLY TRANSPORTATION is no longer a novelty; it has become a necessity in addressing the ecological challenges we face. By embracing electric vehicles, public transportation, cycling and walking, carpooling, sustainable aviation, and alternative fuels, we can significantly reduce our carbon footprint while simultaneously reaping individual health and economic benefits. The importance of collaboration between governments, industries, and individuals cannot be overstated in achieving a greener and more sustainable future of transportation.

1. Green Transport Solutions

Green Transport Solutions

Green transport solutions refer to transportation methods and practices that prioritize environmental sustainability and reduce harmful emissions. With the increasing concerns over climate change and the harmful effects of traditional transportation systems, the development and adoption of green transport solutions have become crucial. This article aims to explore the various aspects of green transport solutions, including their benefits, types, and challenges.

Benefits of Green Transport Solutions

1. Environmental Sustainability: Green transport solutions significantly reduce the carbon footprint and other pollutants associated with transportation. By transitioning to electric or hybrid vehicles, utilizing renewable fuels, and implementing sustainable practices, greenhouse gas emissions and air pollution can be greatly reduced.

2. Improved Air Quality: Traditional transportation systems heavily rely on fossil fuels, leading to the emission of pollutants such as nitrogen oxides and particulate matter. By utilizing green transport solutions, these harmful emissions can be minimized, resulting in improved air quality, particularly in urban areas where pollution is a major concern.

3. Energy Efficiency: Green transport solutions often focus on maximizing fuel efficiency or utilizing alternative energy sources. This helps to reduce overall energy consumption and dependence on finite resources, contributing to a more sustainable and resilient transport system.

4. Health Benefits: The reduction in air pollution resulting from green transport solutions can have significant health benefits, such as reduced respiratory and cardiovascular diseases. Additionally, active transportation options like walking and cycling promote physical activity, leading to improved overall health and well-being.

Types of Green Transport Solutions

1. Electric Vehicles (EVs): Electric vehicles are becoming increasingly popular as an eco-friendly alternative to traditional gasoline-powered vehicles. EVs run on electricity stored in rechargeable batteries and produce zero tailpipe emissions. They not only reduce air pollution but also lower noise levels associated with transportation.

2. Hybrid Vehicles: Hybrid vehicles combine an electric motor with a traditional internal combustion engine. They can run on both electricity and gasoline, resulting in reduced fuel consumption and emissions. Hybrid vehicles are an efficient transitional option while the infrastructure for electric vehicles is being expanded.

3. Public Transport Systems: Public transport plays a vital role in green transport solutions as it encourages the use of shared mobility, reducing the number of privately owned vehicles on the road. Governments and local authorities should invest in reliable and well-connected public transport systems to promote greener commuting options.

4. Cycling and Walking Infrastructure: Encouraging walking and cycling as modes of transportation reduces both pollution and traffic congestion. Developing infrastructure for pedestrians and cyclists, such as cycle lanes and footpaths, improves accessibility and safety, thus encouraging active transportation methods.

Challenges Faced by Green Transport Solutions

1. Infrastructure Development: The widespread adoption of green transport solutions requires extensive infrastructure development, such as charging stations for electric vehicles or dedicated bike lanes. Adequate planning and investment are necessary to ensure the infrastructure is in place to support the transition.

2. Cost: Green transport solutions, particularly electric vehicles, often come with a higher upfront cost compared to conventional vehicles. In order to incentivize adoption, governments need to provide financial support, such as tax incentives and subsidies, to make these technologies more affordable for consumers.

3. Range Anxiety: Electric vehicles' limited driving range and lack of charging infrastructure can lead to range anxiety for potential users. Expanding

the charging network and increasing battery technology advancements are essential to address this concern and promote wider electric vehicle adoption.

4. Behavioral Change: Shifting away from traditional transportation habits and rethinking personal mobility choices require a shift in societal attitudes and behaviors. Encouraging people to embrace greener transport options may require education, awareness campaigns, and active public engagement.

Green transport solutions have the potential to revolutionize the way we travel and address the pressing challenges of climate change and air pollution. Although there are challenges to overcome, the benefits of adopting green transport solutions far outweigh the obstacles. With continued investments in infrastructure, financial incentives, and public awareness, we can create a sustainable and efficient transport system that benefits both the environment and human health.

2. Electric Vehicles: The Future of Mobility

Electric vehicles (EVs) are steadily gaining popularity and momentum as the future of mobility. With the rapidly growing concerns surrounding climate change, air pollution, and reliance on fossil fuels, EVs offer a promising solution to address these pressing issues. They provide an environmentally friendly and sustainable alternative to traditional gasoline-powered vehicles, reducing greenhouse gas emissions and improving air quality.

One of the most significant benefits of EVs is their zero tailpipe emissions. Traditional vehicles emit harmful pollutants such as carbon dioxide, nitrogen oxides, and particulate matter, which contribute to air pollution and numerous health problems. EVs, on the other hand, run on electricity stored in batteries and have no tailpipe emissions. This makes them much cleaner and healthier for both the environment and human health.

Additionally, EVs can contribute to significant reductions in greenhouse gas emissions. As the major contributor to climate change, the burning of fossil fuels for transportation releases large amounts of carbon dioxide into the atmosphere. By transitioning to EVs, which rely on electricity generated from renewable sources like solar and wind power, we can significantly reduce carbon emissions and combat global warming.

The advancement of EV technology has also led to improvements in their overall performance and driving experience. Modern electric vehicles are equipped with high-capacity batteries and efficient electric motors, allowing them to offer comparable, and sometimes even superior, performance to their gasoline counterparts. Furthermore, the torque produced by these electric motors provides instant acceleration, resulting in a smooth and powerful driving experience.

In terms of range, EVs have come a long way in recent years. Earlier electric vehicles could only travel limited distances before needing to recharge, leading to what is commonly known as "range anxiety." However, newer EV models now offer extended ranges, with some capable of covering several hundred miles on a single charge. This enhanced range, coupled with the continuous expansion of charging infrastructure, is gradually diminishing range anxiety concerns and making EVs a viable option for long-distance travel.

Furthermore, the cost of owning and operating an electric vehicle is becoming increasingly competitive with traditional internal combustion engine vehicles. While the upfront cost of EVs may still be higher than their gasoline counterparts, this gap is closing as the prices of batteries and other components continue to decline. Moreover, the overall operating costs of EVs are significantly lower. The cost of electricity is generally cheaper than gasoline, and EVs require less maintenance due to fewer moving parts and no need for oil changes.

The electric vehicle revolution is not limited to passenger cars alone. Electric buses and trucks are also gaining traction, offering similar environmental benefits and improved efficiency for public transportation and freight industries. These electrified vehicles in various sectors promise to revolutionize urban transportation, reduce noise pollution, enhance energy resilience, and contribute to sustainable, smart cities.

Governments around the world are recognizing the potential of electric vehicles and implementing supportive policies to accelerate their adoption. Incentives such as tax credits, grants, and subsidies are being offered to incentivize the purchase of EVs and encourage the development of charging infrastructure. Several countries, including Norway, the Netherlands, and China, have set ambitious targets to phase out the production and sales of internal combustion engine vehicles in the near future.

However, there are still challenges to overcome for widespread adoption of electric vehicles. Battery technology needs further advancements to increase energy density and reduce charging times, addressing concerns about limited range and charging infrastructure. Additionally, ensuring a sustainable supply chain for the raw materials used in batteries, such as lithium and cobalt, is crucial to prevent environmental degradation and ethical concerns.

Despite these challenges, the benefits of electric vehicles make them a promising solution for a cleaner, greener, and more sustainable future of mobility. With continued investments in research and development, supportive policies, and technological advancements, electric vehicles have the potential to revolutionize transportation and pave the way towards a carbon-neutral future.

3. Public Transportation and Infrastructure

Public transportation plays a crucial role in our societies by providing a convenient, affordable, and environmentally friendly mode of transportation for a large number of people. It is also a significant component of a city or region's infrastructure, which comprises the physical structures, such as roads, bridges, and public buildings, that support societal activities.

A well-developed public transportation system is a sign of a progressive and advanced society. It offers numerous benefits that enhance the quality of life for individuals and communities. Firstly, public transportation reduces traffic congestion. As the population of cities continues to grow, roads become more congested, leading to increased travel times, frustration, and pollution. According to a study by the American Public Transportation Association, public transportation ridership in the United States saves 4.2 billion gallons of gasoline annually, reducing congestion and cutting the number of miles driven by numerous individuals. Investing in public transportation can alleviate congestion by providing an efficient alternative to private vehicles.

Secondly, public transportation offers cost savings for individuals. Owning a private vehicle incurs expenses such as fuel costs, maintenance, parking fees, and insurance. Public transportation eliminates these costs, providing an economical option for individuals who need to travel regularly or long distances. Affordability makes public transportation attractive to individuals of various socioeconomic backgrounds, thereby promoting social inclusion and equal mobility opportunities for all members of society.

Furthermore, public transportation contributes to a more sustainable future by decreasing pollution and the carbon footprint. Private vehicles are a significant contributor to air pollution, emitting harmful greenhouse gases. According to the Union of Concerned Scientists, the average car produces over 400 grams of CO2 per mile. By encouraging the use of public transportation, cities can significantly reduce emissions, improving air quality and combating climate change. Investing in electric or hybrid vehicles for public transportation further enhances their positive environmental impact.

Public transportation also has a positive impact on public health. Encouraging individuals to use public transportation instead of driving can lead to higher levels of physical activity. Walking or cycling to and from public transport stops not only contributes to daily exercise but also reduces sedentary behavior. Moreover, public transportation can also address transportation challenges faced by vulnerable populations, such as the elderly or disabled, ensuring they have access to essential services and can participate in community activities.

In addition to the benefits for individuals and the environment, public transportation also has economic advantages. A well-planned public transportation system can attract businesses and foster economic development. It provides easy accessibility for employees, customers, and suppliers, enhancing business opportunities. Moreover, public transportation projects create jobs throughout the planning, construction, and operation phases. These jobs range from urban planning and civil engineering to maintenance and administration, injecting capital into local economies and reducing unemployment rates.

Investing in public transportation and infrastructure is vital for governments and policymakers to ensure a well-connected and prosperous society. Effective public transportation systems require significant planning, funding, and continual improvement. Governments should allocate adequate resources for developing and maintaining infrastructure, updating existing transport networks, and introducing new technologies to optimize efficiency and convenience.

In conclusion, public transportation is an essential component of a well-functioning society. It offers numerous benefits, including reduced traffic congestion, cost savings for individuals, environmental sustainability, improved public health, and economic development. Governments and policymakers should prioritize investing in public transportation and infrastructure to ensure safe, efficient, and equitable mobility options for all members of society.

4. Sustainable Freight and Logistics

Sustainable freight and logistics is a crucial aspect of creating a more environmentally friendly and efficient transportation system. It involves the use of sustainable practices and technologies to reduce emissions, waste, and energy consumption in the freight industry. This can help to mitigate climate change, protect natural resources, and improve overall operational efficiency.

There are several key areas where sustainable freight and logistics can make a significant impact. These include transportation mode choice, vehicle efficiency, fuel choices, supply chain optimization, and intelligent transportation systems.

Transportation mode choice is one of the most critical aspects of sustainable freight. By evaluating different modes of transportation such as road, rail, air, and water, companies can make informed decisions that optimize cost, time, and environmental impact. For instance, choosing rail or water transport instead of road transport for long-haul shipments can greatly reduce greenhouse gas emissions.

Vehicle efficiency is another important aspect of sustainable freight. Investments in fuel-efficient vehicles and technologies, such as hybrid, electric, and alternative fuel vehicles, can significantly reduce CO_2 emissions and improve energy efficiency. Efficient vehicle routing and load optimization can also minimize fuel consumption and improve overall productivity.

Fuel choices play a significant role in sustainable freight and logistics. Switching to cleaner fuels, such as natural gas or biodiesel, can significantly lower emissions of pollutants and greenhouse gases. Additionally, advancements in renewable energy sources, such as solar and wind power, can offer alternative energy options for powering vehicles and facilities.

Supply chain optimization is another critical area in sustainable freight and logistics. By streamlining processes, eliminating inefficiencies, and reducing waste, companies can enhance overall operational efficiency and minimize environmental impacts. This can include strategies like just-in-time delivery, consolidation of shipments, and implementing efficient inventory management systems.

Intelligent transportation systems (ITS) also play a vital role in sustainable freight and logistics. These systems utilize advanced technologies, such as GPS, telematics, and data analytics, to optimize transportation networks, improve traffic management, and enhance supply chain visibility. ITS can help reduce congestion, lower fuel consumption, and improve overall logistics efficiency.

Overall, sustainable freight and logistics are crucial for creating a more sustainable and efficient transportation system. By adopting sustainable practices, technologies, and strategies, the freight industry can minimize its environmental footprint, reduce costs, and enhance overall operational performance. This not only benefits businesses but also helps to protect the environment and improve the quality of life for future generations.

Chapter 6: Sustainable Cities and Infrastructure

In recent decades, the rapid pace of urbanization and the increasing demands of modern society have resulted in the growth of cities and the need for efficient and sustainable infrastructure systems. This chapter aims to explore the concepts and strategies behind sustainable cities and infrastructure, focusing on the importance of designing and developing urban areas that can meet the present needs without compromising the ability of future generations to meet their own needs. Through sustainable urban development, cities can become more resilient, livable, and environmentally friendly, fostering a better quality of life for their residents.

1. The Need for Sustainable Cities:

The world is witnessing an unprecedented urban population increase, with over 55% of the global population living in urban areas. This rapid urbanization poses significant challenges in terms of resource depletion, energy consumption, pollution, and social inequality. Sustainable cities provide an effective solution to address these challenges by promoting efficient land use, sustainable transportation systems, and equitable provision of resources and services.

2. Sustainable Urban Planning and Design:

One of the key elements of sustainable urban development is proper planning and design. This involves identifying appropriate locations for different land uses, ensuring mixed land-use patterns, and promoting compact, walkable, and transit-oriented developments. By integrating green spaces, creating pedestrian-friendly environments, and incorporating renewable energy sources, cities can enhance their livability and reduce the environmental impact.

3. Sustainable Infrastructure Systems:

Infrastructure plays a vital role in supporting the functionality and development of cities. Traditional infrastructure systems often consume vast amounts of resources, emit greenhouse gases, and contribute to pollution. Sustainable infrastructure seeks to minimize these negative impacts by adopting renewable energy sources, employing energy-efficient technologies, and implementing innovations in waste and water management. Examples include green buildings, smart grids, and integrated transportation networks.

4. Sustainable Transportation:

Transportation is a significant contributor to urban pollution and energy consumption. To promote sustainability, cities must prioritize efficient and low-carbon transportation options. This includes investing in public transport infrastructure, promoting cycling and walking, and integrating different modes of transportation. By reducing reliance on private vehicles, cities can mitigate congestion, improve air quality, and enhance the overall transportation experience for residents.

5. Resource Efficiency and Waste Management:

Cities consume vast amounts of resources and generate huge volumes of waste. To achieve sustainability, cities must prioritize resource efficiency measures and implement effective waste management strategies. This involves recycling and reuse programs, composting, and minimizing waste generation through improved production methods. By adopting circular economy principles, cities can reduce their environmental footprint and contribute to a more sustainable future.

6. Social Equality and Inclusiveness:

Sustainable cities must not only aim to enhance environmental sustainability but also prioritize social equality and inclusiveness. This means promoting affordable housing, providing equal access to healthcare and education, fostering social cohesion, and addressing urban poverty. Creating inclusive cities ensures that all residents, regardless of their background, benefit from sustainable development initiatives and have an improved quality of life.

DEVELOPING SUSTAINABLE cities and infrastructure is essential for human well-being and the preservation of our planet. By embracing sustainable urban development practices, cities can create a harmonious balance between economic growth, social equality, and environmental protection. The concepts and strategies presented in this chapter offer a glimpse into the tremendous potential of sustainable cities and infrastructure to shape a better future for generations to come.

1. Shaping Sustainable and Resilient Cities

The task of shaping sustainable and resilient cities is a complex and multifaceted one. It requires not only a deep understanding of the environmental, social, and economic challenges that cities face, but also innovative and integrated approaches to address these challenges.

Sustainable cities are those that are designed and developed in a manner that minimizes resource consumption, reduces waste and pollution, and promotes the well-being of its residents. They prioritize the use of renewable energy sources, promote sustainable transportation options, and foster a strong sense of community and social cohesion. Resilient cities, on the other hand, are those that are able to adapt and recover quickly from shocks and stresses such as natural disasters, economic downturns, or social unrest.

One of the key strategies for shaping sustainable and resilient cities is through urban planning and design. Traditional urban planning approaches have often prioritized the needs of cars and developers over those of pedestrians and the environment. This has resulted in cities that are car-centric, sprawling, and fragmented, with a high ecological footprint and a low quality of life for residents. To create sustainable cities, planners and designers need to shift their focus towards creating compact, walkable, and mixed-use developments that promote active transportation, preserve natural spaces, and enhance social interactions.

Another important aspect of shaping sustainable and resilient cities involves the integration of sustainable infrastructure and technologies. This includes the adoption of renewable energy sources, such as solar or wind power, for both residential and commercial buildings. It also includes the implementation of smart technologies that improve energy efficiency, waste management, and emergency response systems. By integrating these technologies into the fabric of the city, we can reduce carbon emissions, enhance resource efficiency, and improve the overall livability and resilience of our urban areas.

Equally important is the need to address social equity and inclusivity in the planning and development of sustainable and resilient cities. Historically, urban planning decisions have often marginalized certain communities or perpetuated socio-economic disparities. To create truly sustainable and resilient cities, we need to ensure that everyone has access to affordable housing, good quality education, healthcare, and public transportation. We also need to involve local communities and stakeholders in the decision-making processes, and empower them to participate in shaping their own neighborhoods.

Lastly, creating sustainable and resilient cities requires a shift in mindset and culture. Individuals, communities, businesses, and governments all have a role to play in creating a sustainable and resilient urban future. This includes adopting sustainable lifestyle choices, such as reducing waste and energy consumption, supporting local businesses, and engaging in sustainable urban agriculture. It also involves fostering a culture of collaboration, innovation, and proactive urban management, where cities constantly evolve and adapt to future challenges.

In conclusion, shaping sustainable and resilient cities is a dynamic and transformative process that requires the integration of various strategies and approaches. By focusing on urban planning, sustainable infrastructure, social equity, and a cultural shift towards sustainability, we can create cities that are not only environmentally-friendly but also vibrant, inclusive, and economically prosperous. The task may be challenging, but the potential rewards for our planet and future generations are immense.

2. Urban Planning for a Greener Future

Urban planning plays a crucial role in shaping cities and ensuring sustainable development. As the world grapples with environmental challenges, such as climate change and dwindling natural resources, it has become imperative to incorporate greener initiatives in urban planning to create a more sustainable future.

One of the key aspects of urban planning for a greener future is transportation. Cities are major contributors to greenhouse gas emissions primarily due to the reliance on automobiles. To combat this, urban planners need to prioritize the development of sustainable transportation networks. This can include promoting the use of public transport, designing walkable communities, and integrating bicycle lanes. By encouraging these modes of transport, urban areas can reduce pollution, ease congestion, and promote a healthier lifestyle for residents.

Another vital aspect to consider in urban planning is green space. Increasing urbanization is resulting in the loss of natural habitats and green areas. However, incorporating parks, gardens, and green corridors into urban plans can have numerous benefits. Green spaces act as havens for wildlife, improve air quality, reduce the urban heat island effect, and provide opportunities for relaxation and recreation. Therefore, urban planning should prioritize the preservation and creation of such spaces, contributing to a healthier and more sustainable urban environment.

Energy efficiency measures should also be incorporated into urban planning for a greener future. Buildings are responsible for a substantial portion of greenhouse gas emissions. Thus, it is essential to design structures that maximize energy efficiency. This can involve implementing green building codes that require energy-efficient designs, materials, and technologies. Additionally, adopting sustainable practices, such as utilizing renewable energy sources like solar power and incorporating passive design strategies, can reduce the environmental footprint of buildings.

Water management is another key consideration in urban planning for a greener future. With increasing urban populations, reliable water supply and efficient wastewater management are crucial. Urban planners need to implement strategies to conserve water, such as promoting efficient irrigation systems, rainwater harvesting, and water-efficient fixtures in buildings. Additionally, incorporating green infrastructure, such as rain gardens and permeable pavements, can also help manage stormwater and reduce runoff, mitigating urban flooding risks.

Furthermore, urban planning for a greener future must also promote social equity and community engagement. Low-income communities often face greater environmental risks and suffer the most significant impacts of climate change. By fostering inclusive urban planning practices, these communities can benefit from greener initiatives, ensuring that everyone has access to clean air, water, and green spaces. Engaging local communities in the planning process gives them a voice and an opportunity to advocate for their needs while contributing valuable insights.

In conclusion, urban planning must play a pivotal role in creating a greener future. By prioritizing sustainable transportation, preserving and creating green spaces, promoting energy efficiency, implementing water management strategies, and fostering social equity, urban areas can become environmentally friendly and sustainable. It is essential for urban planners to collaborate with various stakeholders, including local communities, to ensure that green initiatives are inclusive, cost-effective, and address the unique challenges faced by diverse urban environments. Through thoughtful and vigilant urban planning, cities can become thriving, resilient, and environmentally sustainable hubs for generations to come.

3. Sustainable Buildings and Infrastructure

Sustainable Buildings and Infrastructure

Sustainable buildings and infrastructure are integral components of sustainable development. They contribute towards reducing our carbon footprint and addressing climate change while also improving social and economic well-being. In recent years, there has been a growing emphasis on sustainability in the construction industry, with a shift towards the adoption of green building practices and sustainable design principles.

One key aspect of sustainable buildings is energy efficiency. Energy usage in buildings is one of the major contributors to greenhouse gas emissions. By integrating energy-efficient technologies and strategies, such as solar panels, smart lighting systems, and insulation, buildings can significantly reduce their energy consumption. This not only helps the environment but also results in lower utility bills for occupants.

In addition to energy efficiency, sustainable buildings also focus on water conservation. Water scarcity is a global issue, and buildings account for a substantial portion of water consumption. Sustainable features such as low-flow faucets, rainwater harvesting systems, and greywater recycling help minimize water usage in buildings. These measures not only save water but also reduce the strain on local water resources.

Furthermore, sustainable buildings aim to minimize waste generation and promote recycling. Construction and demolition activities generate a considerable amount of waste, most of which ends up in landfills. By adopting practices like using recycled materials, implementing efficient construction practices, and designing for deconstruction, the construction industry can reduce waste generation and promote a circular economy.

Another critical aspect of sustainable buildings is the selection of materials. Building materials that are renewable, recyclable, locally sourced, and low in embodied energy are preferred. Using sustainable building materials not only minimizes the environmental impact of construction but also promotes the growth of sustainable industries.

Moreover, sustainable buildings emphasize indoor environmental quality. Poor indoor air quality, inadequate natural lighting, and inefficient heating and cooling systems can negatively impact the health and well-being of building occupants. Sustainable buildings focus on providing a healthy and comfortable indoor environment through features like proper ventilation, use of natural daylighting, and efficient HVAC systems.

In addition to sustainable buildings, the development of sustainable infrastructure is essential for achieving sustainability goals. Sustainable infrastructure includes transportation systems, water and wastewater management systems, and energy grids. By integrating sustainable features into infrastructure development, such as bicycle lanes, public transportation networks, green stormwater management systems, and renewable energy installations, we can create greener and more livable cities.

Furthermore, sustainable buildings and infrastructure also promote social and economic sustainability. Collaboration between governments, businesses, and communities can create job opportunities, promote sustainable entrepreneurship, and improve the overall quality of life. For example, the construction of sustainable buildings often requires specialized skill sets, thereby generating new employment opportunities. Additionally, integrating green spaces and public amenities into sustainable infrastructure projects enhances community well-being and creates inclusive spaces for all.

Overall, sustainable buildings and infrastructure play a crucial role in shaping a more sustainable future. By prioritizing energy efficiency, water conservation, waste reduction, material selection, indoor environmental quality, and social and economic well-being, we can mitigate the environmental impact of the built environment while also promoting sustainable development. From reducing greenhouse gas emissions to minimizing waste generation and enhancing community resilience, sustainable buildings and infrastructure hold the key to a more sustainable and resilient future.

(Credit: The above text has been AI-generated)

4. Smart Cities: Utilizing Technology for Sustainability

Smart cities are a growing trend in urban development, as societies look for innovative solutions to tackle the pressing challenges of rapid urbanization, environmental degradation, and resource depletion. These cities utilize cutting-edge technology to not only improve the quality of life for their residents but also ensure sustainability and environmental protection. The concept of smart cities harnesses the power of connected devices, big data, and advanced analytics to streamline services, optimize resource allocation, and reduce energy consumption.

One primary focus of smart cities is on sustainable transportation. These cities integrate various intelligent transportation systems to improve mobility and reduce traffic congestion. For instance, smart traffic management systems employ real-time data from sensors placed on roadways to efficiently manage traffic flow and reduce travel times. This not only improves air quality by reducing emissions but also enhances overall transport efficiency.

Another critical aspect of smart cities is energy management. These cities aim to optimize energy use and promote the integration of renewable energy sources. One approach is the deployment of smart grids, which utilize advanced monitoring and control systems to efficiently distribute electricity. By closely monitoring consumption patterns, smart grids can identify areas of high energy demand and adjust supply accordingly, reducing wastage and enhancing overall efficiency.

Water management is also paramount in smart cities. With water scarcity becoming an increasing concern worldwide, smart cities utilize advanced technologies to ensure efficient water usage and reduce wastage. Smart irrigation systems, for example, employ sensors to monitor soil moisture levels and adjust watering schedules accordingly, reducing water consumption.

Moreover, smart water distribution systems also detect and tackle leaks promptly, minimizing water loss and improving conservation.

In addition to efficient resource management, smart cities prioritize citizen engagement and participation. Technology-enabled platforms such as mobile applications and online portals allow residents to access real-time information about public services, report issues, and contribute suggestions. This not only fosters a sense of community but also enables the effective allocation of resources based on real-time data and citizen insights.

Furthermore, safety and security are crucial considerations in smart cities. By deploying advanced surveillance systems and interconnecting monitoring devices, these cities can effectively improve law enforcement and emergency response mechanisms. For instance, sensors placed in public spaces can detect unusual activities and send immediate alerts to law enforcement agencies, enabling a quick response and preventive measures.

While the concept of smart cities presents numerous benefits, there are also challenges that need to be addressed. Privacy and data security issues are prominent concerns, as the extensive use of technology collects massive amounts of data. Comprehensive policies and regulations must be implemented to ensure that personal information is protected and used ethically.

Moreover, the implementation of smart city technologies requires substantial investment. Governments and private sectors need to work together to secure funding and develop scalable models to make these cities economically viable.

In conclusion, smart cities hold the promise of harnessing technology to build sustainable, efficient, and livable urban spaces. By integrating innovative solutions in transportation, energy, water management, citizen engagement, and security, these cities can solve pressing challenges plaguing urban areas. While there are hurdles to overcome, the potential benefits of smart cities make them an exciting avenue for the future of urban development.

Chapter 7: Green Finance and Investment

In recent years, there has been a significant surge in interest and activity around the concept of green finance and investment. This trend can be attributed to growing concerns about climate change, environmental degradation, and the urgent need to transition towards a sustainable economy. This chapter aims to explore the topic of green finance and investment, providing detailed information on its importance, mechanisms, and potential impact on the global economy.

The Importance of Green Finance and Investment:

Green finance refers to financial services, products, and investments that promote environmental sustainability and tackle climate-related challenges. It plays a crucial role in mobilizing financial resources to support green projects and businesses driving the transition towards a low carbon and resource-efficient economy. The importance of green finance and investment lies in its potential to align financial flows with the goals of sustainable development and facilitate the necessary transition to a greener future.

Green Finance Mechanisms:

There are several mechanisms through which green finance operates. One of the key mechanisms is green bonds, which are fixed income securities issued to fund projects with environmental benefits. These bonds have gained significant traction in recent years, with governments, corporations, and financial institutions tapping into the growing demand for environmentally-friendly investments. Another mechanism is green loans, which are loans specifically designed to finance green projects and initiatives. These loans often come with favorable terms and conditions, incentivizing businesses and individuals to invest in renewable energy, energy efficiency, and other sustainable projects.

Public Sector Initiatives:

The public sector plays a crucial role in driving the green finance and investment agenda. Governments around the world have been implementing various policy measures to promote sustainable finance, such as providing tax incentives, establishing regulatory frameworks, and setting up dedicated institutions to support green initiatives. Additionally, international organizations like the United Nations and the World Bank have been instrumental in fostering global cooperation and facilitating the flow of green finance across borders.

Private Sector Engagement:

The private sector's engagement in green finance and investment is essential for its success and scalability. While there are still challenges in terms of limited awareness and risk perceptions, an increasing number of businesses and investors are recognizing the potential of green finance to generate both financial returns and positive environmental impacts. Furthermore, the integration of environmental, social, and governance (ESG) criteria into investment decisions has gained momentum, with investors increasingly seeking greener and socially responsible opportunities.

Potential Impact on the Global Economy:

The potential impact of green finance and investment on the global economy is immense. By promoting the development and deployment of clean technologies and sustainable practices, it can drive job creation, foster innovation, and enhance competitiveness. Moreover, by redirecting financial resources away from carbon-intensive industries and activities, green finance can contribute to mitigating climate change and reducing environmental risks.

Challenges and Opportunities:

Despite the significant progress made, there are still several challenges that need to be addressed to unlock the full potential of green finance and investment. These include scaling up appropriate financial instruments, setting robust sustainability standards, and aligning financial regulations with sustainable objectives. However, these challenges also present valuable opportunities for governments, financial institutions, businesses, and individual investors to collaborate and accelerate the transition towards a greener and more sustainable economy.

GREEN FINANCE AND INVESTMENT have emerged as powerful tools to address sustainability challenges and unleash the potential of a greener economy. Through their mechanisms and initiatives, they can mobilize financial resources, channel investments towards sustainable projects, and drive the transition towards a low-carbon and environmentally conscious future. As the world grapples with the urgent threats posed by climate change, green finance and investment serve as beacons of hope, offering effective ways to reconcile economic growth with environmental protection.

1. Understanding Green Financing

Green financing refers to the financing of projects and activities that are environmentally-friendly and promote sustainable development. It encompasses a wide range of financial products and services that are specifically targeted towards investments that have positive environmental impacts, such as renewable energy projects, energy efficiency improvements, sustainable agriculture, and climate adaptation projects, among others.

Understanding green financing requires delving into the various types of financial products and services that are available to support investments in these areas. One of the key instruments in green financing is green bonds, which are debt securities issued by governments, municipalities, corporations, and other entities to raise capital for environmentally-friendly projects. Green bonds are similar to traditional bonds, but the proceeds are earmarked for specific green projects and are subject to more stringent reporting and verification requirements to ensure that the funds are used for their intended purpose.

In addition to green bonds, other financial products and services that fall under the umbrella of green financing include green loans, which provide funding for green projects, green mortgages, which offer favorable terms for the purchase or renovation of energy-efficient homes, and green venture capital funds, which invest in early-stage startups that focus on sustainable technologies and business models.

Understanding green financing also requires grasping the concept of impact measurement and reporting. In order to ensure transparency and accountability, projects financed through green instruments often need to demonstrate the positive environmental impact they generate. This can be done through rigorous impact measurement and reporting processes, which assess the potential or actual outcomes of the project in terms of carbon emissions reduction, energy savings, water conservation, waste reduction, or other relevant environmental criteria. These reports help investors and other stakeholders evaluate the effectiveness and sustainability of green investments.

Furthermore, understanding green financing means recognizing the benefits that can be derived from investing in environmentally-friendly projects. Green investments can offer attractive financial returns, as they tap into growing markets and capitalize on the increasing demand for sustainable solutions. Moreover, green projects often enjoy government support and incentives, such as tax breaks, feed-in-tariffs, or subsidy programs, which can enhance the financial viability of these investments. Furthermore, companies and organizations that embrace sustainability and incorporate green practices into their operations can enhance their reputation and attract environmentally conscious consumers and investors.

Lastly, understanding green financing requires acknowledging the challenges and barriers that exist in this field. One of the key challenges is the lack of standardization and harmonization in green finance frameworks and definitions. Different countries and organizations have developed their own criteria and guidelines for what constitutes a green project, which can lead to confusion and hinder cross-border investments. Moreover, the perceived higher risk associated with green projects, such as technology or regulatory risks, can deter some investors from entering this market.

In conclusion, understanding green financing is crucial in harnessing the potential of finance to support environmentally-friendly projects and advance sustainable development. By familiarizing oneself with the various financial products and services that fall under green financing, as well as the measurement and reporting requirements, the benefits, and the challenges associated with this field, individuals and organizations can make informed investment decisions that have positive impacts on the environment.

2. Green Investment Opportunities

Green investment opportunities refer to investment options that focus on environmentally and socially responsible businesses. In recent years, there has been a growing interest in green investments, as environmental concerns have become more prominent and investors seek to align their portfolios with sustainable practices. This article will explore some of the most promising green investment opportunities and provide a comprehensive overview of each.

1. Renewable Energy: Renewable energy is perhaps the most well-known and popular green investment opportunity. It includes investments in solar, wind, hydro, and geothermal energy sources. With a shift towards cleaner and more sustainable energy alternatives, renewable energy investments have seen significant growth in recent years. Investing in renewable energy projects, such as solar farms or wind turbines, allows investors to support the transition to a low-carbon economy.

2. Energy Efficiency: Investments in energy efficiency aim to reduce energy consumption and improve resource management. This can include investments in energy-efficient buildings, smart grid technologies, energy storage solutions, and demand-side management. Energy efficiency investments not only contribute to greenhouse gas emission reductions but also offer attractive financial returns by lowering energy costs for businesses and households.

3. Sustainable Agriculture: This investment opportunity focuses on agricultural practices that promote sustainable land use, reduce water and chemical usage, and protect biodiversity. Sustainable agriculture investments can include funding organic farms, precision agriculture technologies, and agroforestry initiatives. With growing concerns about food security and the impact of conventional farming practices on the environment, sustainable agriculture investments are gaining traction.

4. Waste Management: Investments in waste management involve supporting businesses that promote waste reduction, recycling, and proper disposal. This can include investments in waste-to-energy conversion technologies, recycling facilities, and waste management infrastructure. As governments and companies prioritize circular economy practices and waste reduction targets, waste management investments present attractive opportunities.

5. Green Real Estate: Green real estate investments focus on sustainable building practices, energy-efficient properties, and LEED-certified (Leadership in Energy and Environmental Design) buildings. Investing in green real estate can offer attractive returns while also contributing to reduced energy consumption, water efficiency, and healthier indoor environments. As the demand for environmentally friendly buildings increases, green real estate investments are becoming more prominent.

6. Sustainable Infrastructure: Investments in sustainable infrastructure revolve around funding projects that promote environmentally friendly transportation, communication, and urban planning. This can include investments in electric vehicle charging infrastructure, public transportation systems, telecommunications networks, and sustainable city developments. Sustainable infrastructure investments play a crucial role in reducing greenhouse gas emissions and improving the quality of urban life.

7. Water Treatment and Conservation: This investment opportunity focuses on technologies and businesses that aim to improve water treatment, conservation, and water quality. Investments in water treatment plants, desalination technologies, water-efficient manufacturing processes, and water recycling systems can play a significant role in addressing water scarcity and safeguarding this vital resource for future generations.

In conclusion, green investment opportunities abound, offering investors a chance to align their portfolios with environmentally and socially responsible businesses. From renewable energy and energy efficiency to sustainable agriculture and waste management, there are numerous avenues to invest in a greener future. While these investments can offer financial returns, they also contribute to addressing critical environmental challenges and promoting sustainable development.

3. Financial Institutions and Sustainable Development

Financial institutions play a crucial role in the sustainable development of a country. They are responsible for mobilizing and allocating funds to various sectors of the economy, promoting social and environmental welfare, and facilitating economic growth. By incorporating sustainable development principles into their operations, financial institutions can contribute significantly to a more sustainable future.

One of the primary ways financial institutions promote sustainable development is by financing environmentally friendly projects. For instance, they can provide funding for renewable energy projects such as solar power plants or wind farms. By supporting these initiatives, financial institutions help reduce greenhouse gas emissions, promote clean energy generation, and combat climate change.

Moreover, financial institutions can encourage the adoption of energy-efficient technologies by offering special lending facilities and incentives. SMEs and individuals looking to invest in eco-friendly technologies such as energy-efficient appliances or hybrid vehicles can receive favorable loan terms from banks that prioritize sustainable development.

Financial institutions can also support sustainable development by promoting responsible investing. They can encourage their clients to explore sustainable investment options, such as socially responsible investment (SRI) funds or impact investing. SRI funds invest in companies that adhere to ethical principles and have positive environmental and social footprints. Impact investing involves directing funds towards companies and projects that generate measurable and positive social and environmental impacts while also delivering a financial return. By raising awareness about these investment options, financial institutions empower individuals and organizations to make sustainable investment choices.

Furthermore, financial institutions can establish specific initiatives and programs to address social and environmental issues. For example, they can promote financial literacy programs to educate individuals about sustainable financial planning and investments. These programs can help individuals make informed choices about their finances while considering the long-term impacts of their decisions on the environment and society.

In addition, financial institutions can integrate environmental and social risk assessment into their lending practices. By identifying potential risks, such as environmental pollution or human rights violations, financial institutions can avoid financing projects that could harm the environment or society. This not only protects the institution from reputational risks but also encourages borrowers to adopt sustainable practices to secure funding.

Moreover, financial institutions can work together with governments and NGOs to develop innovative financial products. These products can help address social and environmental challenges, including financing affordable housing projects, promoting entrepreneurship in marginalized communities, and supporting sustainable agriculture. By collaborating with various stakeholders, financial institutions harness collective expertise and resources to achieve sustainable developmental goals more effectively.

Lastly, financial institutions can demonstrate their commitment to sustainable development through transparent reporting and disclosure practices. By disclosing environmental and social performance indicators, financial institutions provide accountability to their stakeholders and drive industry-wide standards. These reports can encourage healthy competition among financial institutions by showcasing their sustainability efforts and driving the adoption of sustainable practices across the industry.

In conclusion, financial institutions have a significant role to play in promoting sustainable development. By focusing on environmentally friendly projects, responsible investment practices, social and environmental risk assessment, and transparent reporting, financial institutions can contribute to a more sustainable future. Their actions have far-reaching impacts on economic development, social well-being, and environmental protection, making them a critical stakeholder in the pursuit of sustainable development goals.

4. Cost-Benefit Analysis of Green Projects

Cost-benefit analysis is a useful tool in evaluating the value and feasibility of various projects. When it comes to green projects, such as those aimed at promoting sustainability and reducing environmental impact, cost-benefit analysis takes on added importance. These projects often involve significant upfront costs but can have long-term benefits for both the environment and the economy. As such, it is important to consider both the costs and the benefits when assessing the viability of green projects.

One of the primary costs associated with green projects is the initial investment required. Whether it's installing solar panels, upgrading to energy-efficient appliances, or implementing recycling programs, these projects can require substantial financial resources. Additionally, the costs of materials, labor, and maintenance need to be taken into account. It is important to properly identify and calculate all costs to get an accurate assessment of the project's financial implications.

On the other hand, green projects offer numerous benefits that should also be considered. Some benefits are direct and quantifiable, such as energy savings, reduced emissions, and lower operational costs. For instance, installing solar panels can lead to long-term electricity savings, which can offset the initial investment. Other benefits, however, are more indirect and harder to measure. These can include improved brand image, increased employee morale, and enhanced community relationships. While these benefits may not have a clear monetary value, they can still have a positive impact on the overall cost-benefit analysis.

In addition to direct and indirect benefits, green projects can also contribute to more sustainable and resilient communities. For example, investing in green infrastructure, such as green roofs or rainwater harvesting systems, can help mitigate the impact of storms and flooding. This can save money on disaster recovery efforts while also providing a safer and healthier environment for residents. Furthermore, green projects can create new job opportunities and stimulate local economies, which can have positive social and economic ripple effects.

To conduct an effective cost-benefit analysis of green projects, it is essential to consider the timeframe over which the costs and benefits will occur. While some benefits may be realized immediately, others may take years to fully materialize. It is important to take a comprehensive and long-term perspective when evaluating the long-term financial viability of green projects.

It is also important to consider potential risks and uncertainties associated with green projects. Potential risks can include changes in regulations or policies, fluctuations in energy prices or market conditions, and technological advancements that may make the project obsolete in the future. These risks should be carefully assessed and factored into the cost-benefit analysis to ensure a realistic and accurate assessment of the project's viability.

In conclusion, cost-benefit analysis plays a vital role in evaluating green projects. By considering both the costs and the benefits associated with these projects, stakeholders can make informed decisions that maximize both environmental and economic outcomes. By properly assessing the financial implications, potential risks, and long-term benefits, cost-benefit analysis provides a valuable framework for evaluating the viability of green projects.

Chapter 8: Green Jobs and Workforce Development

In recent years, the concept of sustainability has gained significant attention worldwide. As individuals, communities, and nations grapple with the consequences of climate change, there is a growing awareness of the need to transition towards a more sustainable and low-carbon future. This transition requires not only significant investment in clean technologies and infrastructure but also a well-prepared and skilled workforce that can drive the growth of green sectors. This chapter explores the concept of green jobs and the importance of workforce development in achieving a sustainable future.

Defining Green Jobs

Green jobs are employment opportunities that contribute to preserving, restoring, or enhancing environmental quality. These jobs span a wide range of sectors, including renewable energy, energy efficiency, waste management, sustainable agriculture, and green construction. The defining characteristic of green jobs is their positive impact on environmental sustainability through the reduction of greenhouse gas emissions, conservation of non-renewable resources, and the promotion of ecosystem health.

Benefits of Green Jobs

The creation of green jobs is believed to have numerous benefits for both the environment and the economy. Firstly, these jobs provide a solution to rising unemployment rates by offering new employment opportunities. Moreover, the growth of green sectors can stimulate economic development by attracting investments, increasing productivity, and driving innovation. This growth can foster the diversification of economies, reduce dependency on fossil fuels, and increase energy security. Additionally, green jobs have the potential to reduce income inequality by offering higher wages and improved working conditions compared to traditional industries.

Workforce Development for Green Jobs

To effectively capitalize on the potential of green jobs, workforce development is crucial. It involves equipping individuals with the necessary skills, knowledge, and capabilities to succeed in green sectors. Such development encompasses education and training programs, as well as job placement initiatives. It also necessitates collaboration between educational institutions, government agencies, industry representatives, and labor unions.

Education and Training Programs

Education plays a fundamental role in preparing individuals for green jobs. Vocational schools, community colleges, and universities should offer specialized programs that focus on renewable energy technologies, sustainable practices, energy conservation, and other relevant subjects. These programs should incorporate both theoretical knowledge and hands-on training to ensure that graduates possess the necessary competencies for green employment. Furthermore, the integration of sustainability principles across various disciplines is essential to cultivate a sustainability-focused mindset among students.

Job Placement Initiatives

Job placement programs aim to connect skilled individuals with employment opportunities in green sectors. They can take the form of internships, apprenticeships, or job referral services. These initiatives should be widely accessible and transparent to ensure that individuals from diverse backgrounds have equal opportunities. Collaboration with industry partners can help tailor these initiatives to the specific needs and demands of the green job market.

Collaboration and Partnerships

Successful workforce development requires collaboration between various stakeholders. Governments at all levels should play a key role in developing policies, providing funding, and coordinating efforts related to green jobs and workforce development. Similarly, industry partners should actively engage in designing curriculum, financing training programs, and providing vocational experiences. Educational institutions must actively seek input from industry representatives when designing their programs to ensure they remain responsive to the evolving needs of the green job market.

THE GROWTH OF GREEN sectors and the establishment of a sustainable future rely on the presence of a well-prepared workforce. Green jobs offer a multitude of benefits, including employment opportunities, economic growth, and environmental protection. By investing in education and training programs and fostering partnerships between stakeholders, countries can develop the necessary workforce to successfully transition towards a low-carbon and sustainable economy. Workforce development for green jobs is not just a policy objective; it is a crucial step towards ensuring a secure and prosperous future for both current and future generations.

1. Overview of Green Employment

Navigating careers in the green sector can be both daunting and exciting. With the growing awareness of environmental issues and the increasing demand for sustainable solutions, there are a plethora of opportunities in this field. However, finding the right path for your individual interests and skills can require some careful consideration.

One of the first steps in navigating a career in the green sector is determining your passion and what specific area you want to focus on. The green sector encompasses a wide range of industries and job roles, including renewable energy, environmental consulting, sustainable agriculture, waste management, and green building, among others. Identifying your area of interest will help you narrow down your job search and make more informed decisions about education and training opportunities.

Once you have identified your area of interest, the next step is to invest in your education and gain the necessary knowledge and skills. Many universities and colleges offer degree programs in environmental sciences, renewable energy, sustainability, and related fields. Consider pursuing a degree that aligns with your career goals or taking specific courses or certifications to gain expertise in your chosen field.

Networking is also crucial in navigating careers in the green sector. Attend industry conferences, join professional organizations, and engage with online communities to connect with professionals already working in the field. Through networking, you can gain valuable insights, learn about new job opportunities, and even find mentors who can guide you in your career.

Internships and volunteer opportunities are another effective way to gain practical experience and make important connections. Many organizations, both large and small, offer internships to students and recent graduates, which can provide valuable hands-on experience and exposure to real-world problems. Volunteering for environmental organizations or participating in community projects can also help you gain experience and demonstrate your commitment to environmental causes.

Additionally, staying current with industry trends and developments is essential in the rapidly evolving green sector. Follow industry publications, attend webinars, and participate in workshops to stay informed about the latest technologies, policies, and initiatives in your chosen field.

Flexibility and adaptability are key qualities needed in the green sector. The field is constantly changing, and new opportunities continue to emerge. Embrace lifelong learning and be open to taking on new challenges and roles to stay competitive in this dynamic industry.

Finally, don't be afraid to take risks and explore unconventional paths. Innovation and creativity are highly valued in the green sector, and thinking outside the box can lead to exciting and rewarding career opportunities. Consider starting your own business or joining a startup that focuses on sustainability or renewable energy. These entrepreneurial ventures can offer you the chance to make a significant impact and shape the future of the industry.

In conclusion, navigating careers in the green sector requires careful consideration, education, networking, experience, and adaptability. Finding your passion, gaining relevant knowledge and skills, networking with professionals, gaining experience through internships and volunteering, staying informed about industry trends, and being open to taking risks are all important steps in forging a successful and fulfilling career in the green sector.

2. Navigating Careers in the Green Sector

Green Skills Development and Training encompasses a wide range of initiatives and programs aimed at equipping individuals with the knowledge and skills needed to thrive in a rapidly changing world. With the increasing focus on sustainability and the transition towards a greener economy, there is a growing demand for professionals and workers who are proficient in green practices and technologies.

One of the key objectives of Green Skills Development and Training is to address the skill gaps that exist in the workforce. Many traditional industries are being disrupted by the transition towards a low-carbon economy, and workers in these industries need to adapt and acquire new skills to remain employable. Furthermore, as green technologies and practices become more prevalent, new job opportunities are emerging, and individuals need to be trained to fill these positions.

Green Skills Development and Training can take various forms, depending on the needs and requirements of different stakeholders. For individuals, this can involve attending workshops, courses, and training programs to acquire specific green skills relevant to their field of expertise. For example, engineers may need training on renewable energy systems, architects may need to learn about sustainable design principles, and entrepreneurs may need to understand the green business landscape.

For businesses, Green Skills Development and Training can focus on upskilling existing employees or hiring individuals with the necessary green skills. This can involve partnering with educational institutions or training providers to develop customized programs that align with the company's sustainability goals and objectives. By investing in green skills development, businesses can not only stay ahead of regulatory requirements but also gain a competitive edge in a market increasingly driven by sustainability considerations.

From a broader perspective, Green Skills Development and Training plays a crucial role in driving the transition towards a sustainable economy. The knowledge and skills acquired by individuals through green training programs enable them to contribute to environmental conservation, resource management, and the development of innovative solutions to environmental challenges.

Moreover, green skills are also transferable skills that can be applied beyond specific industries or sectors. For example, skills such as critical thinking, problem-solving, and collaboration are essential in addressing complex environmental issues. Green training programs often encompass these broader skills, ensuring that individuals are equipped with a holistic skill set that can be valuable in various contexts.

In conclusion, Green Skills Development and Training is vital for individuals, businesses, and society as a whole. It provides individuals with the tools and knowledge necessary to navigate the transition towards a sustainable economy and equips businesses with the workforce needed to thrive in a changing landscape. By investing in green skills development, we can ensure a sustainable future for all.

3. Green Skills Development and Training

Promoting inclusive and decent green jobs is crucial in order to achieve a sustainable and equitable future. Green jobs refer to employment opportunities that contribute to preserving or restoring the environment while also benefiting individuals and society as a whole. They can be found in a wide range of sectors, including renewable energy, energy efficiency, sustainable agriculture, waste management, and green infrastructure.

Inclusivity is a key aspect of promoting green jobs. It entails ensuring that these opportunities are accessible to all individuals, regardless of their gender, age, ethnicity, or socioeconomic background. In many countries, women and marginalized groups are often underrepresented in the workforce, particularly in high-skilled roles. Therefore, efforts should be made to promote inclusivity in the green job sector by providing training and education programs that cater to the needs and capabilities of these groups.

Creating and promoting decent green jobs is equally important. Decent employment refers to work that guarantees fair wages, good working conditions, job security, and social protection. Green jobs should not only aim to protect the environment but also enhance the quality of life for workers. This can be achieved by ensuring fair remuneration, job stability, opportunities for career growth and continuous learning, and the provision of comprehensive social benefits.

In order to effectively promote inclusive and decent green jobs, governments, businesses, and civil society organizations need to work together. Policymakers play a crucial role in developing supportive frameworks and policies that encourage the creation of green jobs and ensure their inclusivity and decent working conditions.

Investing in green education and training programs is a critical step towards promoting inclusive and decent green jobs. These programs should focus on equipping individuals with the necessary skills and knowledge to thrive in green sectors. Additionally, promoting gender-responsive policies and initiatives that address gender inequalities in the workforce can help increase the representation of women in green jobs.

Businesses can also contribute by adopting green practices and investing in sustainable and inclusive processes within their operations. Providing equal employment opportunities, implementing fair labor standards, and supporting diversity and inclusion initiatives are some ways in which businesses can promote inclusive and decent green jobs.

Moreover, civil society organizations and NGOs can play an important role in advocating for the rights of workers, pushing for policy changes, and creating platforms for dialogue and collaboration between stakeholders. They can also engage in capacity-building activities to empower marginalized groups to participate in the green job sector.

In conclusion, promoting inclusive and decent green jobs is essential for building a sustainable and equitable future. Inclusivity ensures that everyone has equal access to these opportunities, while decency guarantees fair wages and good working conditions. To achieve this, governments, businesses, and civil society organizations must collaborate and invest in education, training, and policies that support green sectors and prioritize the well-being of workers. By doing so, we can foster a more sustainable and inclusive economy that benefits both people and the planet.

4. Promoting Inclusive and Decent Green Jobs

Chapter 9: Sustainable Consumption and Production

SUSTAINABLE CONSUMPTION and production are crucial components of achieving a sustainable future. This chapter delves into the concept, challenges, and strategies related to sustainable consumption and production. It seeks to provide a comprehensive understanding of the topic and inspire readers to make more informed and sustainable choices in their everyday lives.

What is Sustainable Consumption and Production?

Sustainable consumption and production can be defined as the use of goods and services that minimize the negative impacts on natural resources, the environment, and human well-being. It involves reducing waste, optimizing resource use, and promoting the longevity of products and their efficient use.

The Challenges of Unsustainable Consumption and Production

Unsustainable consumption and production practices have significant negative consequences for the environment, society, and future generations. These challenges include resource depletion, pollution, greenhouse gas emissions, and worsening social and economic inequalities. Rapid population growth and unsustainable consumption patterns further exacerbate these challenges.

Strategies for Sustainable Consumption and Production

There are several strategies that individuals, businesses, and governments can adopt to promote sustainable consumption and production:

1. Education and Awareness: Increasing public knowledge about sustainable consumption and production is critical to driving behavioral change. Schools and educational institutions should incorporate sustainability into their curriculum, and others can raise awareness through campaigns, media, and community engagement.

2. Resource Efficiency: Optimizing the use of resources through efficient processes, recycling, and waste reduction is essential for sustainable consumption and production. This involves promoting circular economy approaches, where materials are recycled and reused rather than disposed of.

3. Sustainable Products and Services: Encouraging the production and consumption of sustainable products is pivotal in achieving sustainable consumption and production goals. This includes environmentally friendly products and services, such as energy-efficient appliances, organic foods, and sustainably sourced materials.

4. Life Cycle Assessment: Conducting a life cycle assessment of products and services enables the identification of their environmental impacts throughout their lifespan. This information allows consumers to make more informed choices and businesses to evaluate and improve their processes.

5. Extended Producer Responsibility: Holding producers responsible for the entire life cycle of their products, including recycling and disposal, is crucial to reducing waste and promoting sustainable consumption and production. Governments can enforce regulations and incentivize producers to adopt sustainable practices.

6. Sustainable Public Procurement: Governments and public institutions can lead by example through sustainable procurement practices. By prioritizing sustainable products and services, they can encourage the market to shift towards sustainability.

7. Collaborative Initiatives: Partnerships between businesses, governments, NGOs, and consumers are pivotal in driving change. Collaboration enables the sharing of best practices, expertise, and resources to implement sustainable consumption and production strategies effectively.

SUSTAINABLE CONSUMPTION and production are essential for building a sustainable future. By adopting strategies such as education, resource efficiency, sustainable products and services, life cycle assessment, extended producer responsibility, sustainable public procurement, and collaborative initiatives, we can minimize negative environmental impacts, promote social equity, and secure the well-being of future generations. It is up to individuals, businesses, and policymakers to take action and embrace sustainable consumption and production practices for a more sustainable world.

Note: The information provided in this chapter is based on scholarly research and expert opinions, making it a valuable and insightful resource for readers interested in the topic of sustainable consumption and production.

Chapter 9: Sustainable Consumption and Production

Introduction to Sustainable Consumption

Sustainable consumption refers to the practice of using resources and goods in a manner that minimizes adverse ecological and social impacts, while promoting a better quality of life for all. It recognizes the interconnectedness of our consumption patterns and their impact on the environment, society, and the economy.

The concept of sustainable consumption has gained prominence in recent years as global challenges such as climate change, resource depletion, and social inequality have become more apparent. It emphasizes the need for individuals, businesses, and governments to adopt a more responsible and ethical approach towards consumption.

There are several key principles that underpin sustainable consumption. These include reducing overall consumption levels, using resources more efficiently, prioritizing renewable and sustainable sources of energy, promoting fair trade and ethical production practices, and minimizing waste and pollution.

One of the main goals of sustainable consumption is to minimize the ecological footprint of our daily activities. This footprint includes the amount of land, water, and resources required to produce and consume goods and services. By reducing our consumption levels, we can reduce the strain on these finite resources, and ensure their availability for future generations.

Another important aspect of sustainable consumption is the promotion of social justice. It recognizes that the benefits and burdens of consumption are not evenly distributed, and seeks to ensure fair and equitable access to goods and services for all. This means considering the needs and rights of marginalized groups, as well as addressing issues such as poverty, inequality, and exploitation in global supply chains.

Sustainable consumption also entails a shift towards more sustainable and ethical production practices. This includes promoting the use of renewable energy sources, reducing emissions and pollution, eliminating hazardous substances, and promoting the use of eco-friendly materials. It also involves supporting businesses and industries that prioritize the welfare of workers, respect human rights, and promote gender equality.

Individuals play a crucial role in promoting sustainable consumption. By making mindful choices about what we consume and how we consume, we can contribute to the wider shift towards a more sustainable and equitable society. This can range from opting for greener modes of transportation, consuming less meat and dairy, buying second-hand or locally-sourced products, reducing waste and recycling, and supporting businesses that prioritize sustainability and social responsibility.

Achieving sustainable consumption requires a collective effort from individuals, businesses, and governments. While individuals can make a significant impact through their everyday choices, businesses need to take responsibility for the sustainable production and marketing of goods and services. They can do this by adopting sustainable practices, engaging in eco-design, promoting recycling and resource recovery, and investing in renewable energy. Governments also have a crucial role to play in setting regulations and incentives that promote sustainable consumption, as well as supporting research and development in green technologies.

In conclusion, sustainable consumption is an important concept that seeks to address the environmental, social, and economic challenges of our time. It recognizes the need for a more responsible and ethical approach towards consumption, and emphasizes the importance of reducing waste, conserving resources, and promoting social justice. By adopting sustainable consumption practices, we can work towards a more sustainable and equitable future for all.

1. Introduction to Sustainable Consumption

Consumer behavior refers to the study of individuals, groups, or organizations and the processes they use to select, secure, use, and dispose of products, services, experiences, or ideas to satisfy their needs and wants. In recent years, there has been a growing interest and importance placed on green choices, both from consumers and businesses alike.

Green choices, also known as environmentally friendly or sustainable choices, refer to the decisions made by individuals and organizations that minimize harm to the environment or promote its preservation. These choices can include anything from buying products made from renewable materials to using energy-efficient appliances or choosing to walk or bike instead of drive.

One reason why consumer behavior and green choices have become intertwined is the increasing awareness of environmental issues. With rising concerns over climate change, deforestation, pollution, and other environmental problems, consumers are becoming more conscious of the impact their choices have on the planet.

Additionally, consumers' attitudes towards the environment are evolving. They no longer view green choices as just a means to make them feel good about themselves or to be part of a trendy movement. Consumers now recognize that their choices have real consequences and can make a difference in the overall health of the planet.

Moreover, consumer behavior can also be influenced by external factors such as government regulations and incentives. For instance, policies that promote the use of renewable energy or impose penalties for carbon emissions can sway consumers towards greener choices. Similarly, offering tax credits or subsidies for eco-friendly products can incentivize consumers to opt for greener alternatives.

The availability and affordability of green choices also play a significant role in consumer behavior. Historically, environmentally friendly products, such as organic foods or energy-efficient appliances, were seen as more expensive and limited in availability. However, with advancements in technology and increased demand, these products are becoming more accessible and affordable for the average consumer. As a result, more people are choosing green options simply because they are now valued for their quality and affordability.

Social influence is another important factor in consumer behavior and green choices. People are often influenced by the opinions and behaviors of those around them. For example, if friends, family, or influential celebrities are endorsing and embracing green choices, it can motivate others to do the same. This phenomenon is often seen in social media, where influencers and celebrities promote eco-friendly products or lifestyles, influencing their followers to make greener choices.

Furthermore, businesses are recognizing the demand for green products and are adjusting their marketing strategies accordingly. They are tapping into consumers' growing interest in sustainability by highlighting the environmental benefits of their products. This strategy is not only aimed at attracting customers who actively seek out green choices but also at enticing those who would not normally prioritize environmental considerations. It's a win-win situation for businesses and consumers alike, as companies benefit from increased sales while consumers have more green choices available to them.

In conclusion, the study of consumer behavior and green choices unveils interesting insights into how individuals make decisions and the factors that influence those decisions. With the ongoing concern for the environment, consumers are increasingly valuing and prioritizing green choices. Whether driven by personal values or external incentives, the impact of consumer behavior on the environment continues to shape the marketplace and influence businesses to provide more eco-friendly options.

2. Consumer Behavior and Green Choices

Sustainable product design and manufacturing is a comprehensive approach that focuses on developing products and their associated processes in an environmentally and socially responsible manner. It aims to minimize the negative impacts on the environment and human health throughout the product lifecycle, including resource extraction, manufacturing, distribution, use, and end-of-life disposal.

One of the key principles of sustainable product design is the consideration of environmental factors during the product development stage. Designers need to evaluate the entire lifecycle of the product and identify opportunities for improvement. This includes the selection of materials and processes that have low environmental impact, as well as the minimization of waste and pollution.

A fundamental aspect of sustainable product design is the use of renewable and recyclable materials. Designers should prioritize the use of materials that can be easily recycled or biodegraded, as this reduces waste and promotes a circular economy. Additionally, the sourcing of materials should be done in an ethical manner, considering factors such as social and environmental responsibility in the supply chain.

Another important aspect of sustainable product design is energy efficiency. Designers need to optimize the energy consumption of products by minimizing electricity usage, employing energy-saving technologies, and incorporating renewable energy sources where possible. This not only reduces the carbon footprint of products but also decreases energy costs for users.

Furthermore, sustainable product design involves the creation of products that are durable and easily repairable. By designing products that can be repaired and upgraded, the lifespan of the product is extended, reducing the need for frequent replacements. This helps to conserve resources and reduce waste generation.

Manufacturing processes also play a crucial role in sustainable product design and manufacturing. A sustainable approach involves the efficient use of resources such as raw materials, water, and energy. Manufacturers need to implement practices that minimize waste generation, enhance resource conservation, and utilize eco-friendly technologies.

Additionally, sustainable manufacturing often involves the adoption of cleaner production methods. This includes the utilization of non-toxic materials, implementing pollution prevention strategies, and reducing emissions and effluents. Advanced technologies like 3D printing can also be employed to minimize waste generation in the manufacturing process.

Furthermore, sustainable product design and manufacturing also involves the consideration of social factors. This includes ensuring fair working conditions for workers involved in the production process, as well as supporting local communities and promoting social responsibility. Ethical practices and socially responsible approaches are important for maintaining a sustainable business model.

In conclusion, sustainable product design and manufacturing is a holistic approach that takes into account environmental, economic, and social aspects. By integrating sustainability principles into product development and manufacturing processes, we can create products that are not only environmentally friendly but also meet the needs of users and contribute to a more sustainable future.

3. Sustainable Product Design and Manufacturing

In recent years, there has been a growing trend of retailers implementing sustainable practices as part of their business models. This shift towards sustainability is driven by various factors, including consumer demand, environmental concerns, and the potential for cost savings and brand enhancement. Retailers have recognized that sustainability is not only beneficial for the planet but also for their bottom line. As a result, many have taken proactive steps to reduce their environmental footprint and promote green practices throughout the entire supply chain.

One way in which retailers are driving sustainability is by adopting energy-efficient measures in their stores and distribution centers. This includes the installation of energy-saving lighting systems, the use of solar panels, and the implementation of smart building technologies that optimize energy consumption. By reducing energy usage, retailers not only decrease their carbon emissions but also lower their operational costs, making it a win-win situation. Moreover, retailers are also encouraging their suppliers to adopt energy-efficient practices and reduce their carbon footprint.

Another key area where retailers are making significant strides in sustainability is in waste management. Many retailers have implemented recycling programs in their stores, encouraging customers to dispose of their waste in an environmentally friendly manner. From providing recycling bins for paper, plastic, and glass to incentivizing customers with discount or loyalty program points, retailers are actively engaging customers to participate in waste reduction efforts.

Furthermore, retailers are making conscious efforts to reduce packaging waste, which has been a significant contributor to environmental pollution. This involves partnering with suppliers to develop packaging solutions that are more sustainable, using recycled materials, and minimizing excess packaging.

Some retailers have even gone a step further by offering packaging-free options, where customers can bring their containers to fill their products, reducing the need for single-use plastic packaging altogether.

An important aspect of sustainability for retailers is ethical sourcing. Many retailers are now actively working towards ensuring their products are sourced from suppliers who meet strict environmental and social responsibility criteria. This includes working with suppliers that adhere to fair trade practices, provide safe working conditions for employees, and promote sustainable farming techniques. By doing so, retailers are not only supporting sustainable supply chains but also contributing to the well-being of workers and local communities.

Interestingly, retailers are also using sustainability as a tool to attract and retain customers. Consumers, especially millennials and Gen Z, are increasingly conscious of their environmental impact and are actively seeking out retailers that align with their values. A survey conducted by Cone Communications found that 87% of consumers are more likely to purchase products from a company that supports social or environmental issues. Consequently, retailers are leveraging sustainability as a marketing tool, promoting their eco-friendly practices, and providing transparency about their sustainability initiatives in order to enhance their brand image and attract more consumers.

In conclusion, retailers are playing a major role in driving sustainability. Their efforts in adopting energy-efficient measures, implementing recycling programs, reducing packaging waste, and practicing ethical sourcing are not only benefiting the environment but also positively impacting their bottom line. Moreover, by promoting sustainability as a marketing tool, retailers are able to attract and retain customers who are increasingly concerned about the impact of their purchasing decisions on the planet. With the growing importance of sustainability in the retail industry, it is expected that more retailers will continue to pave the way for a greener future.

4. Retailers Driving Sustainability

Chapter 10: Corporate Social Responsibility in a Green Economy- A detailed exploration of an emerging paradigm

IN RECENT YEARS, THE concept of Corporate Social Responsibility (CSR) has gained significant traction as society grapples with the implications of a rapidly changing world and mounting concerns over environmental degradation. The emergence of a "green economy" has intensified the call for businesses to recognize their responsibility towards the environment and society at large. Chapter 10 delves into the multifaceted aspects of CSR in the context of a green economy, offering a comprehensive and engaging discussion on this growing trend.

1. Defining the green economy:

The opening of the chapter provides a comprehensive definition and understanding of what constitutes a green economy. It elucidates the interconnectedness between economic growth, environmental sustainability, and social welfare, highlighting the importance of adopting a holistic approach that considers the well-being of both people and the planet.

2. CSR in a green economy:

The subsequent sections delve into the fundamental relevance of CSR in addressing the intertwined challenges posed by environmental degradation and social inequality in a green economy. Detailed insights are provided on the role that businesses can play in promoting sustainable practices, reducing carbon emissions, conserving resources, and supporting local communities.

3. Leveraging green technologies and innovations:

One of the key strengths of Chapter 10 is its in-depth exploration of how businesses can leverage green technologies and innovations to drive sustainable growth while fulfilling their CSR obligations. This includes a comprehensive analysis of renewable energy solutions, sustainable supply chains, green manufacturing processes, and the benefits these initiatives can bring to the economy, society, and the environment.

4. The regulatory landscape and CSR:

Chapter 10 recognizes the role of government regulation in shaping CSR practices within a green economy. It investigates the various regulations and policies that compel businesses to integrate environmentally-friendly practices and outlines how companies can adapt to these changing legislative frameworks.

5. Stakeholder engagement and collaboration:

An intriguing facet of this chapter lies in its emphasis on stakeholder engagement and collaborative efforts in the realm of CSR. The narrative highlights the importance of companies working with communities, NGOs, and other stakeholders to address their concerns and foster a shared responsibility towards sustainable development, effectively positioning businesses as catalysts for social change.

6. Measuring and reporting CSR impact:

Understanding the measurement and reporting of CSR initiatives is essential for companies committed to transparency and accountability. Chapter 10 comprehensively explores different frameworks, methodologies, and metrics commonly used to measure and communicate the impact of CSR initiatives, offering practical guidance for businesses that aim to assess and report on their sustainability efforts accurately.

CHAPTER 10 PROVIDES a wealth of information and thought-provoking perspectives on the integration of CSR within a green economy. The nuanced analysis of each topic, the wealth of data presented, and the captivating writing style make for a comprehensive and engaging reading experience. By showcasing the potential benefits of incorporating CSR into a green economy, this chapter leaves readers with the conviction that businesses have a crucial role to play in shaping a sustainable and equitable future.

Chapter 10: Corporate Social Responsibility in a Green Economy

In today's rapidly changing world, sustainability has become a key factor for businesses to consider. Companies that take into account social, environmental, and economic factors are not only able to reduce their negative impact on the planet, but also create long-term value for themselves and society.

Integrating sustainability into business strategies requires a multi-faceted approach. It involves analyzing the impacts of the business across its operations, supply chain, and product life cycle. By identifying the areas where the company can improve its environmental performance, it can set goals and develop action plans to achieve a more sustainable future.

One way businesses can integrate sustainability is by implementing efficient waste management practices. This includes reducing, recycling, and properly disposing of waste to minimize its impact on the environment. Companies can engage in innovative recycling initiatives or explore circular economy models that promote resource efficiency and reduce waste generation.

Energy conservation is another crucial aspect of sustainability. By implementing energy-efficient technologies and processes, businesses can not only reduce their carbon footprint but also save costs in the long run. This can involve investing in renewable energy sources like solar panels or exploring energy-saving practices such as the use of energy-efficient appliances and machinery.

In addition, businesses can consider the social aspect of sustainability by investing in human capital. By developing training programs and providing opportunities for employee growth and development, companies can foster a positive and inclusive work environment. This can improve employee satisfaction and loyalty, leading to increased productivity and ultimately, better financial performance.

Furthermore, supply chain management plays a vital role in integrating sustainability into business strategies. Companies can work closely with their suppliers to ensure ethical sourcing and reduce environmental impacts. This may involve choosing suppliers who adhere to sustainable business practices or developing codes of conduct that suppliers must follow. By promoting sustainability throughout the supply chain, businesses can effectively manage risks, enhance their reputation, and contribute to the well-being of communities in which they operate.

Finally, businesses must also consider the economic aspect of sustainability. While integrating sustainability may require an upfront investment, the long-term benefits often outweigh the costs. Sustainable practices can lead to improved efficiency, reduced resource consumption, and increased innovation. They can also help businesses stay ahead of emerging regulations and customer expectations. Companies that proactively integrate sustainability into their strategies are better positioned to withstand future challenges and stay competitive in the ever-evolving global market.

To effectively integrate sustainability into business strategies, companies need to have a holistic and long-term mindset. It should be incorporated into all aspects of the business, from strategy development to product design, marketing, and stakeholder engagement. By doing so, businesses can create value for themselves and society, while preserving the planet for future generations.

1. Integrating Sustainability into Business Strategies

Voluntary initiatives for sustainable development have gained significant attention in recent years as businesses and organizations strive to minimize their environmental impact, enhance social equity, and foster economic growth. These initiatives, spearheaded by both individual companies and collaborative partnerships, offer innovative solutions and best practices that go beyond regulatory compliance.

One such example of a voluntary initiative is the United Nations Global Compact (UNGC). The UNGC encourages businesses to align their strategies and operations with ten universally accepted principles in the areas of human rights, labor, environment, and anti-corruption. By signing on to the UNGC, companies commit to integrating these principles into their practices and reporting progress in achieving them. This initiative emphasizes the importance of corporate citizenship and promotes sustainable development goals (SDGs) as outlined by the United Nations.

Another noteworthy voluntary initiative is the Carbon Disclosure Project (CDP), which acts as a platform for companies to disclose their greenhouse gas emissions and climate change strategies. By encouraging transparency and accountability, the CDP enables investors and stakeholders to make informed decisions and encourages companies to adopt practices that reduce their carbon footprint. Through the CDP, businesses are proactively addressing climate change while also improving operational efficiency and reducing costs.

Private sector-led initiatives, such as the Roundtable on Sustainable Palm Oil (RSPO), illustrate how collaboration among industry players can drive sustainable practices. The RSPO sets and certifies standards for the production, sourcing, and use of palm oil, encouraging responsible cultivation while minimizing environmental and social impacts. By complying with these standards, companies contribute to preserving natural resources, protecting biodiversity, and safeguarding the rights of workers and local communities.

Some voluntary initiatives focus on more specific areas, such as water stewardship. For instance, the CEO Water Mandate is a private-public initiative that mobilizes business leaders to address water-related challenges. Signatories commit to responsible water practices throughout their operations and supply chains, collaborating on projects that improve water availability and quality. This initiative is particularly significant in water-scarce regions, where businesses play a crucial role in securing this vital resource for sustainable development.

Other voluntary initiatives, such as fair trade certifications and responsible sourcing guidelines, directly promote social and economic development. Fairtrade International ensures that farmers and producers in developing countries receive fair prices, decent working conditions, and environmental sustainability. Similarly, the Responsible Jewellery Council (RJC) sets standards for responsible business practices in the jewelry supply chain, enhancing transparency and ensuring ethical sourcing and production.

Voluntary initiatives for sustainable development not only drive positive change but also provide businesses with a competitive edge in the global market. Consumers increasingly prefer companies that demonstrate their commitment to social and environmental responsibility, leading to improved brand reputation and customer loyalty. These initiatives also create opportunities for collaboration and knowledge sharing. By joining forces, businesses can pool resources, expertise, and funding to address complex sustainability challenges that are beyond the scope of individual efforts.

However, it is important to note that voluntary initiatives should not replace robust regulatory frameworks. Government regulations and policies play a crucial role in driving sustainability at both national and international levels.

In conclusion, voluntary initiatives for sustainable development are integral to creating a more inclusive, equitable, and environmentally responsible future. They empower businesses and organizations to go beyond compliance and proactively address global challenges. By adopting these initiatives, companies can contribute to the achievement of sustainable development goals, enhance their brand reputation, and align with growing consumer demands for social and environmental responsibility.

2. Voluntary Initiatives for Sustainable Development

Corporate social responsibility (CSR) is a philosophy that has gained significant prominence in recent years. It refers to a company's commitment to operating in an economically, socially, and environmentally responsible manner.

Implementing CSR initiatives is not only beneficial for society but can also bring long-term advantages to companies. However, it is crucial to follow best practices and ensure that CSR efforts are authentic and impactful.

One of the fundamental best practices in CSR is conducting a thorough assessment of a company's social and environmental impact. This assessment helps identify the areas where the company can make a difference and allows for the alignment of CSR initiatives with organizational values and goals. By conducting this assessment, companies can determine how they can integrate CSR into their core business strategies.

Transparency is another essential aspect of effective CSR. Companies need to communicate their CSR initiatives to stakeholders, including employees, customers, and the general public, in a clear and concise manner. Transparency builds trust and helps demonstrate the authenticity of CSR efforts. Companies should also disclose key metrics and performance indicators regularly to show their progress and accountability.

Collaboration and partnerships are key to successful CSR programs. Collaborating with other organizations, both private and public, NGOs, and governments allows for more extensive and impactful initiatives. For instance, partnering with organizations specializing in social or environmental issues can bring specialized knowledge and expertise, enhancing the effectiveness of CSR efforts.

Having measurable goals and objectives helps in evaluating the success of CSR initiatives. Establishing key performance indicators (KPIs) allows companies to track their progress and make necessary adjustments. KPIs can include metrics such as greenhouse gas emissions reduction, employee volunteer hours, or the number of disadvantaged individuals served. Measuring and reporting on these indicators not only helps assess impact but also demonstrates the company's commitment to its CSR efforts.

Companies should also ensure employee engagement in CSR initiatives. By involving employees in volunteering activities or giving them opportunities to provide input on CSR strategies, companies foster a sense of purpose and pride within the workforce. Engaged employees can act as ambassadors for CSR efforts, positively impacting the organization's reputation and culture.

Finally, regular evaluation and continuous improvement are essential for maintaining effective CSR practices. Companies should periodically review their CSR strategies, identify areas of improvement, and adapt their approaches as societal needs and expectations evolve. Ongoing evaluation helps ensure that CSR initiatives remain relevant, meaningful, and aligned with the changing dynamics of society.

In conclusion, embracing best practices in CSR is vital for companies aiming to play a positive role in society while also reaping long-term benefits. Conducting thorough assessments, promoting transparency, fostering collaboration, setting measurable goals, engaging employees, and continuously evaluating and improving efforts are all key elements of effective CSR initiatives. By following these practices, companies can demonstrate their commitment to social responsibility and create a lasting positive impact on the world.

3. Best Practices in Corporate Social Responsibility

Accountability and reporting play a crucial role in green operations. As businesses worldwide become more environmentally conscious, it is essential for them to track and report their progress in implementing sustainable practices. Accountability ensures that companies are responsible for their environmental impact, while reporting allows for transparency and comparability between different organizations. This article will delve into the importance of accountability and reporting in green operations and highlight some key points to consider.

First and foremost, accountability holds businesses responsible for their actions and their impact on the environment. By establishing clear goals and targets, companies can be held accountable for meeting specific objectives related to sustainability. This accountability can come from both internal and external sources. Internally, businesses can set up robust systems to monitor and evaluate their environmental performance, ensuring that they are meeting their own expectations. Externally, companies may be subject to regulations and reporting requirements imposed by governmental bodies or industry associations. This external accountability ensures that companies are meeting legal and industry standards.

Accountability can take various forms in green operations. One common approach is to designate specific individuals or teams responsible for sustainability initiatives within an organization. These individuals should have the authority and resources needed to implement green policies effectively. Furthermore, setting up clear metrics and systems for tracking progress along the sustainability journey is crucial. These metrics can include CO_2 emissions reductions, waste management practices, water conservation efforts, and more. By setting specific targets and regularly monitoring performance, businesses can demonstrate their commitment to sustainable practices.

Similarly, reporting is integral to green operations as it allows businesses to share their environmental efforts and progress with stakeholders. Reporting can take the form of formal sustainability reports, which outline an organization's sustainability strategy, goals, and achievements over a specific period. These reports are often developed following guidelines provided by recognized frameworks, such as the Global Reporting Initiative (GRI) or the Sustainability Accounting Standards Board (SASB). By adhering to these frameworks, companies ensure that their reports are credible and comparable.

Sustainability reports provide a comprehensive overview of a company's environmental impacts, disclosing key performance indicators (KPIs), and highlighting areas of success and areas for improvement. This transparency allows for benchmarking and comparison between different organizations, promoting healthy competition and driving continual improvement. Additionally, sustainable reporting enhances a company's reputation and credibility, as it demonstrates their commitment to environmental responsibility.

When it comes to accountability and reporting in green operations, it is crucial to consider the broader context and societal expectations. The public is increasingly demanding environmental responsibility from businesses, with consumers favoring brands that prioritize sustainability. Therefore, it is not only important to set goals and report on progress but also to actively engage with stakeholders and directly communicate about sustainability initiatives. Companies that actively communicate their efforts can build trust and strengthen their relationship with customers, employees, investors, and the wider society.

In conclusion, accountability and reporting are essential components of green operations. Accountability ensures that businesses take responsibility for their environmental impact, while reporting provides transparency and comparability for stakeholders. By setting goals, tracking progress, and reporting on sustainability initiatives, companies can demonstrate their commitment to environmental responsibility, enhance their reputation, and build trust with stakeholders. Furthermore, integrating sustainability into business practices can drive innovation, cost savings, and long-term success. Overall, accountability and reporting in green operations are essential for businesses striving to create a more sustainable future.

4. Accountability and Reporting in Green Operations

In any organization, accountability and reporting are crucial components of maintaining green operations. Green operations refer to the practices and strategies that companies adopt to minimize their environmental impact and promote sustainability.

When it comes to accountability, organizations need to establish clear roles and responsibilities within their operations. This includes designating specific individuals or departments responsible for monitoring and ensuring the implementation of green practices. By clearly defining these roles, everyone within the organization understands their responsibilities towards green initiatives. This fosters a sense of accountability among employees, encouraging them to comply with sustainability measures.

Moreover, organizations must establish mechanisms to measure and evaluate their adherence to green practices. This involves setting performance indicators and targets related to environmental impact. For example, a manufacturing company may measure its carbon emissions or waste production and set reduction targets over a specific time frame. By regularly tracking and reporting on these indicators, organizations can monitor their progress towards sustainability goals and identify areas where improvements are necessary.

Accurate reporting is a crucial tool for organizations to communicate their commitment to green operations to stakeholders. Organizations can achieve this through the establishment of robust reporting practices. This includes collecting and analyzing data on environmental performance, and preparing regular reports that provide information on their progress. These reports may focus on important aspects such as energy consumption, greenhouse gas emissions, waste management, and water usage.

Transparency in reporting is also essential to build trust and credibility with stakeholders. Organizations should make these reports publicly available, ensuring that they are easily accessible to employees, customers, investors, and the wider community. By demonstrating transparency, organizations can show their commitment to environmental responsibility and engage stakeholders in their sustainability journey.

It is also critical for organizations to integrate proper reporting systems into their operations. This includes investing in effective software tools and systems that can accurately collect, analyze, and report data related to green practices. With these systems in place, organizations can streamline their reporting processes, reduce the possibility of errors, and ensure that their reports are accurate and reliable.

Moreover, reporting on green operations is not limited to external stakeholders; it also plays a crucial role in fostering internal awareness and accountability. By sharing regular reports on environmental performance with employees, organizations can raise awareness about the impact of individual actions on sustainability. This empowers employees to take ownership of their role in promoting green operations and motivates them to contribute actively to sustainability efforts.

To conclude, accountability and reporting are essential in maintaining green operations in organizations. Through clear accountability mechanisms, organizations can ensure that individuals are held responsible for implementing and monitoring green practices. Reporting, on the other hand, allows organizations to communicate their commitment to sustainability, build trust with stakeholders, and identify areas for improvement. By integrating these practices into their operations, organizations can make significant strides towards a greener and more sustainable future.

Conclusion. The Wider Impacts of the Green Economy Transformation

In our modern world, where environmental concerns have become increasingly important, carving a path to a sustainable future has become a top priority. As humans, we hold the power to shape our future, and it is crucial that we make thoughtful choices and take deliberate actions to ensure a sustainable and thriving planet for generations to come. This article will delve into the intricacies of carving our path to a sustainable future, exploring different aspects and suggestions for a more sustainable world.

First and foremost, it is essential to understand the concept of sustainability. Sustainability refers to meeting our present needs without compromising the ability of future generations to meet their own needs. It is a delicate balance between environment, society, and economy, where all three must be considered and fostered simultaneously. Achieving sustainability requires comprehensive planning, which encompasses a range of elements: from energy and resource consumption to waste management and beyond.

One crucial aspect of carving our path to a sustainable future is transitioning to clean and renewable sources of energy. Fossil fuels have proven to be detrimental to our environment, contributing to greenhouse gas emissions and climate change. Therefore, seeking alternatives such as solar, wind, hydro, and geothermal energy is imperative. Investing in research and development of renewable technologies and encouraging their widespread use will significantly reduce our carbon footprint and lessen our impact on the Earth's delicate ecosystems.

Another factor to consider is the responsible management of resources. With our growing population and increasing consumption patterns, it is more important than ever to conserve resources and use them efficiently. This requires implementing sustainable agriculture practices to reduce water usage and minimize the use of harmful chemicals. Additionally, promoting circular economy models that focus on recycling, reusing, and reducing waste will help minimize our overall resource consumption and create a more sustainable future.

Furthermore, building sustainable communities is key to carving our path to a sustainable future. This involves designing cities and neighborhoods that prioritize walkability, public transportation, and green spaces. By doing so, we can reduce our reliance on cars and decrease pollution levels, while also fostering social connectivity and improving overall well-being. Additionally, sustainable urban planning should incorporate energy-efficient buildings, with proper insulation, lighting, and renewable energy integration.

Education is also crucial in our journey towards sustainability. Teaching the younger generations about the importance of conservation, environmental stewardship, and sustainable practices will ensure a sustainable future. By integrating sustainability into the curriculum and educating individuals from an early age, we can foster a conscious mindset and create a new generation of environmentally aware individuals who will actively contribute to a sustainable world.

In conclusion, carving our path to a sustainable future requires comprehensive and collaborative efforts from individuals, communities, businesses, and governments alike. Transitioning to clean and renewable sources of energy, responsible resource management, sustainable community development, and education are all essential elements to consider. By committing to these actions and prioritizing long-term ecological well-being over short-term gains, we can shape a world where both humans and the environment can thrive harmoniously. Together, we possess the power to carve a path to a sustainable future and leave a positive legacy for future generations.

2. Carving Our Path to a Sustainable Future

In today's world, the concept of sustainability has gained significant attention. With issues like climate change, deforestation, and pollution becoming increasingly dire, it has become evident that we need to find new ways to approach how we live and interact with our environment. While this may seem like a daunting task, carving a path to a sustainable future is not only necessary but also achievable.

One of the key aspects of sustainability is reducing our carbon footprint. This entails finding cleaner and more efficient ways to produce and consume energy. Over the years, there has been a significant shift towards renewable sources of energy such as solar and wind power. These renewable technologies not only produce electricity without emitting harmful greenhouse gases, but they also have the potential to be more cost-effective in the long run.

Another crucial step towards sustainability is the conservation and responsible use of resources such as water. With increasing population and economic growth, the demand for water has skyrocketed. This puts pressure on already scarce resources, but through innovative approaches like rainwater harvesting and better water management systems, we can ensure that everyone has access to clean water without overexploiting our natural sources.

Furthermore, sustainable agriculture practices are vital in addressing food security and climate change. Traditional farming methods often involve harmful pesticides, excessive water use, and deforestation. Transitioning to organic farming, agroforestry, and regenerative farming techniques promotes soil health, preserves biodiversity, and reduces the need for synthetic inputs. By implementing such practices on a global scale, we can not only ensure abundant and nutritious food for future generations but also combat climate change by sequestering carbon in the soil.

Transportation also plays a significant role in carving our path to a sustainable future. The transition to electric vehicles and the development of efficient public transportation systems are key in reducing carbon emissions from the transport sector. Additionally, promoting active modes of transportation like cycling and walking can not only contribute to reducing our carbon footprint but also improve public health by encouraging physical activity.

However, savings will come primarily from reducing the amount of waste generated and improving waste management systems. The concept of the circular economy, where products are designed for durability, repairability, and recyclability, aims to minimize waste and create a closed loop system. Proper waste management involves recycling, composting, and waste-to-energy initiatives to ensure that precious resources are not lost or sent to landfills.

Education and awareness also play a crucial role in creating a sustainable future. It is important to educate individuals from a young age about the importance of sustainability and how their actions can make a difference. By instilling environmentally conscious habits and fostering a mindset of responsibility towards the planet, we can ensure that future generations understand and actively work towards a sustainable future.

In conclusion, carving our path to a sustainable future requires a comprehensive approach encompassing various aspects of our daily lives. By transitioning to renewable energy, conserving water, adopting sustainable agriculture practices, promoting clean transportation, implementing proper waste management systems, and educating individuals, we can move closer to creating a world that sustains both human needs and the health of our planet. Though the journey may be challenging, the rewards are boundless, with a future filled with clean air, water, and a bountiful and vibrant natural environment.